Finite Rings with Identity

PURE AND APPLIED MATHEMATICS

A Series of Monographs and Textbooks

1. K. Yano. Integral Formulas in Riemannian Geometry (1970)
2. S. Kobayashi. Hyperbolic Manifolds and Holomorphic Mappings (1970)
3. V. S. Vladimirov. Equations of Mathematical Physics (A. Jeffrey, editor; A. Littlewood, translator) (1970)
4. B. N. Pshenichnyi. Necessary Conditions for an Extremum (L. Neustadt, translation editor; K. Makowski, translator) (1971)
5. L. Narici, E. Beckenstein, and G. Bachman, Functional Analysis and Valuation Theory (1971)
6. D. S. Passman. Infinite Group Rings (1971)
7. L. Dornhoff. Group Representation Theory (in two parts). Part A: Ordinary Representation Theory. Part B: Modular Representation Theory (1971, 1972)
8. W. Boothby and G. L. Weiss (eds.). Symmetric Spaces: Short Courses Presented at Washington University (1972)
9. Y. Matsushima. Differentiable Manifolds (E. T. Kobayashi, translator) (1972)
10. L. E. Ward, Jr. Topology: An Outline for a First Course (1972)
11. A. Babakhanian. Cohomological Methods in Group Theory (1972)
12. R. Gilmer. Multiplicative Ideal Theory (1972)
13. J. Yeh. Stochastic Processes and the Wiener Integral (1973)
14. J. Barros-Neto. Introduction to the Theory of Distributions (1973)
15. R. Larsen. Functional Analysis: An Introduction (1973)
16. K. Yano and S. Ishihara. Tangent and Cotangent Bundles: Differential Geometry (1973)
17. C. Procesi. Rings with Polynomial Identities (1973)
18. R. Hermann. Geometry, Physics, and Systems (1973)
19. N. R. Wallach. Harmonic Analysis on Homogeneous Spaces (1973)
20. J. Dieudonné. Introduction to the Theory of Formal Groups (1973)
21. I. Vaisman. Cohomology and Differential Forms (1973)
22. B.-Y. Chen. Geometry of Submanifolds (1973)
23. M. Marcus. Finite Dimensional Multilinear Algebra (in two parts) (1973)
24. R. Larsen. Banach Algebras: An Introduction (1973)
25. R. O. Kujala and A. L. Vitter (eds.). Value Distribution Theory: Part A; Part B, Deficit and Bezout Estimates by Wilhelm Stoll (1973)
26. K. B. Stolarsky. Algebraic Numbers and Diophantine Approximation (1974)
27. A. R. Magid. The Separable Galois Theory of Commutative Rings (1974)
28. B. R. McDonald. Finite Rings with Identity (1974)

Finite Rings with Identity

BERNARD R. McDONALD
Department of Mathematics
University of Oklahoma
Norman, Oklahoma

MARCEL DEKKER, INC. New York

MARCEL DEKKER, INC.

270 Madison Avenue, New York, New York 10016

LIBRARY OF CONGRESS CATALOG CARD NUMBER: 74-19018

ISBN: 0-8247-6161-8

Current printing (last digit):

10 9 8 7 6 5 4 3 2 1

PRINTED IN THE UNITED STATES OF AMERICA

To B. M. Stewart,

a teacher and gentleman.

CONTENTS

=152

PREFACE

This monograph arose from a collection of lectures on topics in finite rings at The University of Oklahoma. The motivation for preparing this survey was three-fold: (I) it was hoped that finite rings could provide a wealth of concrete and instructive examples for students mastering the first principles of ring theory and that this monograph could provide a supplementary text for a course in general ring theory; (II) for algebraists, this text provides a survey of the present state of the theory of finite rings with identity; (III) to provide a source material for the growing areas of discrete algebraic modeling. In our perspective the third reason was the greatest contributing factor in the preparation of these notes. Research on discrete linear modeling over finite fields is notably extensive and presently expanding, producing results in combinatorics, finite geometries, algebraic cryptography, linear coding theory, finite linear sequential machines, etc. Questions are now being posed concerning models over more general coefficient rings, for example, Hjelmslev planes, or models which are themselves finite rings, for example, cyclic linear codes. However, often a person working in linear modeling is not extensively trained in algebra. Thus it was hoped we could provide a survey of the theory of finite rings with identity accessible to a reader who has had a beginning course in algebra at the 'Herstein' level.

There are three notable exclusions in this monograph. First, we examined only finite (associative) rings with identity. There was already a treatment of finite nilpotent rings by Kruse and Price [17] and thus our monograph was thought to complement theirs. Second, we concentrated on questions of structure theory and did not examine enumeration and combinatorial problems except for a few exercises. For example, [17] and [119]-[121] examine estimates and order of magnitude bounds on the number of finite rings of a given order. The literature abounds with enumeration problems related to structures over finite fields. Third, we did not provide a treatment of the Galois theory of an arbitrary finite ring. It was thought that the automorphism theory of general finite rings was not as complete as that available for finite commutative rings. The interested reader may examine the references [164]-[179] for results in this direction.

We generally examined four aspects of finite rings: the internal structure of a ring, the structure of modules over the ring, the structure of the group of units of a ring, and the structure of the group of automorphisms of the ring. Chapters (I) through (XII) provide an introduction to the basic theory of finite rings and their modules. Chapters (XIII) through (XVIII) examine finite commutative rings. A propose of these sections was to show that the Galois theory of finite commutative local rings is as sharp as that well-known for finite fields. Chapters (XIX) through (XXI) examine further topics in the general theory of finite rings by an approach which exploits the theory of Galois rings.

Now comes the very pleasant task of thanking the people and groups who made this work possible. Edwin Clark, Joel Brawley and John Fulton made valuable suggestions. Gary Ganske, Leon McQueen, Jim Pomfret and Richard Wirt, who as graduate students suffered willingly through many lectures over these topics, contributed much of their own imagination and work to the finished product. Kirby Smith read the manuscript and offered corrections and improvements. We are indebted to The University of Oklahoma Research Institute for support during the summer of 1972, when most of the manuscript was written and also to both the Graduate Dean, Art Gentile, and the Arts and Sciences Dean, Paige Mulhollan, for support during the summer of 1974. The Chairman, Gene Levy, of the O.U. Mathematics Department has provided assistance and advice. The first ten chapters were typed by the Faculty Research typists. The remaining chapters were typed by Trish Abolins. Mrs. Abolins also cheerfully typed and retyped all the preliminary versions and thankfully still remains my friend. Finally, a word of thanks to my wife, Jean, for her continuing encouragement.

Finite Rings with Identity

I. INTRODUCTION AND ELEMENTARY RESULTS

In this introductory section we observe that the group decomposition of the additive group of a finite ring into Abelian p-groups induces a ring direct sum decomposition. Thus, for most problems in finite rings, we may assume the ring has cardinality p^t and characteristic p^s for some prime p .

Throughout these notes the term "ring" will mean a finite ring with identity. Customarily the identity is denoted by 1 .

WARNING: There are four exceptions to the assumption that "ring" implies a finite number of elements. These are (1) algebraic closures of finite fields, (2) the rational integers Z , (3) polynomial rings and (4) skew-polynomial rings.

Let R denote a ring and let S and T denote subsets of R . The smallest subring of R which contains S is called the ring generated by S and denoted by $\langle S\rangle$ We let $S + T$ denote $\{s + t \mid s \in S, t \in T\}$ and ST denote $\{\sum s_i t_i \mid s_i \in S, t_i \in T\}$. The characteristic of S denoted by $\chi(S)$ is the least positive integer n such that $nx = 0$ for all x in S . The additive group of R is denoted by R^+ , the units of R by R^*, the cardinality of S by $|S|$ and if S is a subring the index of S^+ in R^+ is denoted by $[R : S]$.

We assume the reader has a knowledge of elementary algebra and finite Abelian groups. We let $\sigma(n)$ denote a cyclic group of order n and the type of an Abelian p-group will be denoted by $[n_1,\ldots,n_k]$.

The group R^+ of R decomposes into primary components. It is easy to see that this group decomposition induces a ring decomposition,

<u>(I.1)</u> <u>THEOREM</u>. A ring R is uniquely expressible as a direct sum of rings of prime power order.

Thus, if R is a ring

$$R \simeq \oplus \sum_p R_p \qquad \text{(as rings)}$$

where for each prime p

$$R_p = \{x \in R \mid p^n x = 0 \text{ for some } n \geq 1\}.$$

It is clear that if S is a subring (ideal, left ideal, right ideal) then

$$S \simeq \oplus \sum_p (S \cap R_p)$$

and

$$R/S \simeq \oplus \sum_p (R_p/R_p \cap S).$$

Thus whenever convenient we restrict our attention to rings R such that for some prime p and integers m and n we have $\chi(R) = p^m$

and $|R| = p^n$. In this case, if Z denotes the rational integers, the ring generated by 1 in R is isomorphic to Z/Zp^m. Thus we customarily identify $\langle 1 \rangle$ with Z/Zp^m.

If R is a ring we denote the <u>center</u> of R by $Z(R) = \{r \mid rx = xr$ for all x in $R\}$.

A ring R is <u>decomposable</u> if R is a direct sum of rings R_1 and R_2, $R = R_1 \oplus R_2$. If R is not decomposable, R is called <u>indecomposable</u>.

<u>(I.1)</u> <u>EXERCISE</u>. Prove Theorem (I.1) and the comments following (I.1).

<u>(I.2)</u> <u>EXERCISE</u>. Show that if a ring R is indecomposable then $|R|$ is a power of a prime. If p is a prime show that Z/Zp^n is indecomposable.

<u>(I.3)</u> <u>EXERCISE</u>. Let R be a ring and x be an element of R. Show there is a unique ring morphism

$$\phi_x : Z[X] \to R$$

given by $1 \to 1$ and $X \to x$. Deduce that $\langle 1,x \rangle$ is a commutative subring of R. From this, show

(a) If $|R| = \chi(R)$, then $R \simeq Z/Z\chi(R)$.

(b) If $|R| = p^2$ (p a prime), then R is commutative.

(c) If $|R|$ is cube-free then R is commutative.

(I.4) EXERCISE. Show that $Z/Z4$, $(Z/Z2)[X]/(X^2)$, $(Z/Z2) \oplus (Z/Z2)$ and $(Z/Z2)[X]/(X^2 + 1)$ are all the rings of 4 elements.

(I.5) EXERCISE. Let p be a prime. Show that if R is a non-commutative ring and $|R| = p^3$ then

$$R \simeq \left\{ \begin{bmatrix} a & b \\ 0 & c \end{bmatrix} \mid a,b,c, \text{ in } Z/Zp \right\}.$$

(I.6) EXERCISE. (Euler phi-function) Recall the Euler phi-function $\phi : Z \to Z$ (on the positive integers) is defined as follows:

$\phi(1) = 1$,

If $n > 1$, $\phi(n)$ is the number of integers t, $1 \leq t < n$ and $(t,n) = 1$.

Show

(a) If $n = p^t$ (p a prime) then
$\phi(n) = p^t - p^{t-1} = p^t(1 - \frac{1}{p})$.

(b) If $(r,s) = 1$ then $\phi(rs) = \phi(r)\phi(s)$.

(c) If $p_1,\ldots,p_m$ are the distinct prime factors of n then
$\phi(n) = n(1 - \frac{1}{p_1}) \ldots (1 - \frac{1}{p_m})$.

(d) $\sum_{d|n} \phi(d) = n$ (where the summation extends over positive divisors of n).

(I.7) EXERCISE. Let R^* denote the units of a ring R.

(a) Show that $\phi(n) = |(Z/Zn)^*|$.

(b) Show that $(Z/Zn)^*$ is a cyclic group if $n = 2$, 4, or p^r (p odd prime).

(c) Show that if $(Z/Zn)^*$ is cyclic then $(Z/Zn)^*$ has $\phi(\phi(n))$ generators.

(I.8) EXERCISE. (Ganesan [115]) Let R be a commutative ring. Recall if $a \neq 0$ and b in R satisfy $ab = 0$ then b is a zero divisor (0 is the trivial zero divisor).

(a) Show that if a is not a zero divisor then a is a unit.

(b) Show that if R has only n non-zero divisors then $|R| \leq (n + 1)^2$.

(c) Suppose that if $|R| = m \neq p^a$ (p prime) then show R must contain at least n non-zero divisors where $n = \sqrt{m} - 1$ (if m is square) or n is the greatest integer $< \sqrt{m}$ (if m is not a square).

(d) Show that Z/Zp^2 (p prime) has exactly p zero divisors and p units.

(e) Show that R is a finite field if R has no non-trivial zero divisors.

(I.9) EXERCISE. (Koh [122]) An element a of a ring R is a left zero divisor if there is a $b \neq 0$ in R with $ba = 0$. If $n \geq 1$ and R has only $n + 1$ left zero divisors, show $|R| \leq (n + 1)^2$.

(I.10) EXERCISE. Let $R_1,\ldots,R_n$ be rings and $R = R_1 \oplus \ldots \oplus R_n$.

(a) Show $R^* = R_1^* \times \ldots \times R_n^*$ (direct product of groups).

(b) Show $R[X] = \oplus \sum_{i=1}^{n} R_i[X]$ ($R[X]$ denotes the polynomial ring in one commuting indeterminate over R).

(c) Show $M_m(R) = \oplus \sum_{i=1}^{n} M_m(R_i)$ ($M_m(R)$ denotes the $m \times m$ matrix ring over R).

(I.11) EXERCISE. For a positive integer n, let $R(n)$ be a complete set of representatives of the isomorphism classes of rings of order n, and let $\rho(n) = |R(n)|$. If $n = p_1^{a_1} \ldots p_t^{a_t}$ is the prime factorization of n, show

$$\rho(n) = \rho(p_1^{a_1}) \ldots \rho(p_t^{a_t}) .$$

(I.12) EXERCISE. (Szele) Let R be a ring. Suppose the Abelian group R^+ has minimal basis $\{m_1,\ldots,m_t\}$. Then R is determined by describing the products of basis elements

$$m_i m_j = \sum_s a_{isj} m_s$$

where a_{isj} are integers. This equation allows the introduction of $t \times t$ matrices $A^{(j)}$ where $A_{is}^{(j)} = a_{isj}$ and where the i-th column is taken modulo the order of m_i. If $|R| = p^s$ (p prime), show that R is isomorphic to a ring of $t \times t$ matrices whose i-th columns are interpreted modulo $p^{v(i)}$ where $p^{v(i)}$ is the order of m_i.

II. FINITE FIELDS

Finite fields enter the theory of finite rings in numerous locations and, though they are best known finite rings, the research in finite fields and structures over finite fields is rich and continuing. Much of the motivation of these notes involved attempts to sharpen results known for rings by requiring the rings be finite - analogous to the situations well-known for fields.

Section (A) sketches the standard theory of finite fields and their polynomial rings - for additional detail the reader may refer to any introductory algebra text. Section (B) provides a proof that finite division rings are finite fields. Sections (C) and (D) examine the skew-polynomial ring over a finite field (which will be used later in obtaining results on the structure of finite rings) and more specialized results on the subspaces in finite fields. Section

(D) presupposes a knowledge of basic linear algebra.

Section (A). Finite Fields and Their Polynomial Rings.

The theory of finite rings is intimately connected with finite fields. We recall in this section the basic facts concerning finite fields.

We let Z denote the ring of rational integers and the coset $a + Zm$ of the quotient ring Z/Zm will generally be denoted only by a . If k is a finite field then its characteristic is $\chi(k) = p$ a prime and k contains Z/Zp . Let $\bar{k}$ denote the algebraic closure of k . (See Ex. (II.19).)

(II.1) THEOREM. Let k be a finite field with $\chi(k) = p$. Then

(a) The cardinality $|k| = p^n$ where $n = [k : Z/Zp]$ the degree of k over Z/Zp .

(b) The units k^* of k form a cyclic multiplicative group of order $p^n - 1$.

(c) If k_1 is a subfield of k then k is a simple extension of k_1 . Indeed, a generator of k^* will generate k over k_1 .

(d) For each natural number n , set

$$k_n = \{a \in \overline{Z/Zp} \mid a^{p^n} - a = 0\} .$$

Then k_n is a field of p^n elements, k_n is the only field of p^n elements in $\overline{Z/Zp}$, and $k \simeq k_n$.

(e) Let k_1 be a subfield of k. Then k is separable over k_1 and if a is in k then the irreducible polynomial $\mathrm{Irr}(a,k_1)$ of a over k_1 is a factor of $X^{p^n} - X$ in $k_1[X]$.

Proof. For (a) note k is a Z/Zp-vector space of Z/Zp-dimension $n = [k : Z/Zp]$. To show (b) recall a polynomial f over a field may have at most $\deg(f)$ distinct zeros in the field. Since k^* is an Abelian group with $|k^*| = p^n - 1$, if a is in k^*, a satisfies $X^{p^n - 1} = 1$. Also, since k^* is an Abelian group,

$k^* = \oplus \prod_{i=1}^{t} \sigma(d_i)$ where $d_i \mid d_{i+1}$ for $1 \leq i \leq t - 1$. Hence $p^n - 1 = d_1 \ldots d_t$ and every element of k^* satisfies $X^{d_t} = 1$. We have $d_t \leq p^n - 1 \leq d_t$. Hence $t = 1$ and k^* is cyclic. Now (c) is easy.

Every subfield of p^n elements of $\overline{Z/Zp}$ is contained in $k_n = \{a \in \overline{Z/Zp} \mid a^{p^n} - a = 0\}$. Hence, viewing k naturally as a subfield of $\overline{Z/Zp}$, $k \subseteq k_n$. Hence, to show (d), we need only show $|k_n| = p^n$ and that k_n is a field. The formal derivative of $X^{p^n} - X$ is -1 thus $X^{p^n} - X$ is a separable polynomial and has distinct zeros in $\overline{Z/Zp}$. Thus $|k_n| = p^n$. Obviously if a and b are in k_n then ab is in k_n. Further since p divides

$\binom{p^n}{t}$, $0 < t < p^n$, we have $(a + b)^{p^n} = a^{p^n} + b^{p^n}$. Clearly 0 and 1 are in k_n and hence k_n is a finite domain in $\overline{Z/Zp}$ and thus a finite field.

Finally (e) follows since if g is a polynomial in $k_1[X]$ then $g(a) = 0$ if and only if $Irr(a,k_1)$ divides g .

For a prime p the above field k_n is denoted by $GF(q)$, $q = p^n$, and is called the Galois Field of q elements. Then if k is a finite field we identify $\bar{k}$ with $\overline{Z/Zp}$ and k with a suitable k_n in $\overline{Z/Zp}$.

(II.2) THEOREM. Let $q = p^t$ for a prime p .

(a) In $GF(q)[X]$ there exist irreducible polynomials of degree n for each natural number n .

(b) The field $GF(q^n)$ is the splitting field for all irreducible polynomials of degree n in $GF(q)[X]$.

(c) Let g be an irreducible polynomial in $GF(q)[X]$ of degree d . Then g divides $X^{q^n} - X$ if and only if d divides n .

(d) If an irreducible polynomial g in $GF(q)[X]$ has a zero in $GF(q^n)$, then it splits into linear factors in $GF(q^n)[X]$.

(e) The polynomial $X^{q^n} - X$ is a product of all monic irreducible polynomials g in $GF(q)[X]$ with the degree, $deg(g)$, dividing n .

The proofs of the above are straightforward.

Clearly $GF(q)[X]$ is a Euclidean domain and we have the properties of (II.2). Before we examine the automorphisms of a finite field we recall several other properties of $GF(q)[X]$.

Let $q = p^t$, p a prime, and consider $GF(q)[X]$. Let Δ denote the product of the distinct monic irreducible polynomials in $GF(q)[X]$ whose degrees d are proper divisors of n, i.e., d divides n and $1 \leq d < n$. Then

$$\Delta = \text{L.C.M. } \{X^{\tau(u)} - 1\}$$

where $\tau(u) = q^u - 1$ and u ranges over the proper divisors of n. Let g denote $X^{q^n} - X$ (then Δ divides g) and let

$$\Gamma(n,q) = \frac{g}{\Delta}$$

Then $\Gamma(n,q)$ is the product of the distinct monic irreducible polynomials of degree n over $GF(q)$.

Let $n = p_1^{e_1} p_2^{e_2} \cdots p_s^{e_s}$ be the prime factorization of n and $\sum = \{p_1, \ldots, p_s\}$. Let $\sum_t \subseteq \sum$ where $|\sum_t| = t$ and $\sum_t = \{p_{i_1}, \ldots, p_{i_t}\}$ and define the quotient

$$m(\textstyle\sum_t) = \frac{n}{p_{i_1} \cdots p_{i_t}} ,$$

where we set $m(\sum_0) = n$. Note that there exist $\binom{s}{t}$ subsets $\sum_t$ and

$m(\sum_t)$ is a proper divisor of n.

Let $\nu(n,q)$ denote the number of monic irreducible polynomials of degree n in $GF(q)[X]$. Note that $\deg\Gamma(n,q) = n\nu(n,q)$.

For each natural number m define $[m] = X^{q^m} - X$ then $[n] = [m(\sum_0)] = g$. Finally let

$$\lambda_t = \prod_{\sum_t \subseteq \sum} [m(\sum_t)]$$

and

$$\lambda_0 = g .$$

<u>(II.3)</u> <u>THEOREM</u>. (Dedekind's Formula on Irreducible Polynomials) For the above notation:

(a) If $\Delta_0 = \prod_{0\le 2t\le s} \lambda_{2t}$ and $\Delta_1 = \prod_{1\le 2t+1\le s} \lambda_{2t+1}$ then

$$\Gamma(n,q) = \frac{\Delta_0}{\Delta_1} .$$

(b) $\nu(n,q) = \frac{1}{n}[q^n - \sum q^{n/p_1} + \sum q^{n/p_1p_2} - \dots + (-1)^s q^{n/p_1\cdots p_s}]$

which by Möbius inversion (see exercises) gives

$$q^n = \sum_{d|n} d\nu(d,q) .$$

<u>Proof</u>. (A. A. Albert) Since $\lambda_0 = g$ is separable every monic

irreducible polynomial of degree n is a factor of λ_0 and is not a factor of λ_k, $k > 0$.

Let h be an irreducible polynomial divisor of $\Delta_0\Delta_1$ with $\deg(h) = \mu < n$. Then μ divides n properly. Let $q_1,\ldots,q_t$ be those prime factors of n occurring to higher exponent in n than they do in μ.

If $k > t$ then μ does not divide $m(\sum_k)$ and thus h does not divide λ_k for $k > t$. On the other hand, if $k \leq t$ then μ divides $m(\sum_k)$ if and only if $\sum_k \subseteq \sum' = \{q_1,\ldots,q_t\}$.

There exist $\binom{t}{k}$ subsets $\sum_k$ with $\sum_k \subseteq \sum'$, thus h has multiplicity $\binom{t}{k}$ in λ_k. Hence h is a factor of multiplicity

$$\alpha = 1 + \sum_{2\leq 2k\leq t} \binom{t}{2k}$$

of Δ_0 and a factor of multiplicity

$$\beta = \sum_{1\leq 2k+1\leq t} \binom{t}{2k+1}$$

of Δ_1. Then

$$\rho(X) = (1 + X)^t = 1 + \binom{t}{1}X + \cdots + \binom{t}{t}X^t$$

illustrates the fact that α is the sum of the coefficients of the even powers of X and β the sum of the coefficients of the odd powers. Thus $\rho(-1) = \alpha - \beta = (1 - 1)^t = 0$. Hence h has multiplicity 0 in Δ_0/Δ_1 .

Part (b) of the above theorem is standard and occurs in the exercises. The following result is also a brief exercise.

(II.4) THEOREM. Every function $f : GF(q) \to GF(q)$ where $f : \alpha \to \alpha'$ may be represented uniquely as a polynomial of formal degree $q - 1$ by

$$f(X) = \sum_{\alpha \in GF(q)} \alpha'(1 - (X-\alpha)^{q-1}) .$$

Conversely, if R is a finite commutative ring such that every function $R \to R$ is given by a polynomial then R is a finite field.

For elements $\alpha_0, \ldots, \alpha_{n-1}$ in $GF(q)$, the square matrix

$$\begin{bmatrix} \alpha_0 & \alpha_1 & \cdots & \alpha_{n-2} & \alpha_{n-1} \\ \alpha_1 & \alpha_2 & \cdots & \alpha_{n-1} & \alpha_0 \\ \cdot & \cdot & \cdot & \cdot & \cdot \\ \cdot & \cdot & \cdot & \cdot & \cdot \\ \cdot & \cdot & \cdot & \cdot & \cdot \\ \alpha_{n-1} & \alpha_0 & \cdots & \alpha_{n-3} & \alpha_{n-2} \end{bmatrix}$$

is called a cyclic matrix.

(II.5) THEOREM. (König-Rados) The number of distinct non-zero roots of $f(X) = \sum_{i=0}^{q-2} \alpha_i X^i$ in $GF(q)[X]$ is

$$q - 1 - r$$

where r is the rank of the cyclic matrix of coefficients.

<u>Proof</u>. Let $\beta_1, \ldots, \beta_{q-1}$ denote the non-zero elements of $GF(q)$. Let A be the cyclic matrix determined by $\alpha_0, \ldots, \alpha_{q-2}$ and let

$$B = \begin{bmatrix} 1 & \beta_1 & \cdots & \beta_1^{q-2} \\ \vdots & & \ddots & \vdots \\ 1 & \beta_{q-1} & & \beta_{q-1}^{q-2} \end{bmatrix}.$$

Observe $\beta_i^{q-1} = 1$. Then

$$BA = \begin{bmatrix} f(\beta_1) & \beta_1^{-1} f(\beta_1) & \cdots & \beta_1^{-(q-2)} f(\beta_1) \\ \vdots & & & \\ f(\beta_{q-1}) & \beta_{q-1}^{-1} f(\beta_{q-1}) & \cdots & \beta_{q-1}^{-(q-2)} f(\beta_{q-1}) \end{bmatrix}.$$

Let N denote the number of roots of f, i.e., the number of the zeros in $\{f(\beta_1), \ldots, f(\beta_{q-1})\}$. Without loss we may assume the zeros are

$$\beta_{q-1-N}, \ldots, \beta_{q-1}.$$

Then the rank of BA is less than or equal to $q - 1 - N$. Indeed, it is exactly $q - 1 - N$ since a principal submatrix of this order has as its determinant the Vandermond determinant of $\beta_1^{-1}, \ldots, \beta_{q-1-N}^{-1}$ multiplied by $f(\beta_1) \cdots f(\beta_{q-1})$.

If L is an extension field of K, let $G_K(L)$ denote the group of K-field automorphisms of L. Let $\overline{GF(q)}$ denote the algebraic closure of $GF(q)$ where $q = p^n$ (p a prime). The mapping

$$\Theta : \overline{GF(q)} \to \overline{GF(q)}$$

$$\Theta : x \to x^q$$

is a field automorphism called the <u>Frobenius automorphism</u> of $\overline{GF(q)}$.

<u>(II.6)</u> <u>THEOREM.</u>

(a) The Frobenius automorphism $\Theta : \overline{GF(q)} \to \overline{GF(q)}$ satisfies the following:

(i) If k is a field with $GF(q) \subseteq k \subseteq \overline{GF(q)}$, then Θ is a $GF(q)$-automorphism of k.

(ii) $GF(q^s) = \{a \text{ in } \overline{GF(q)} \mid \Theta^s(a) = a\}$

(iii) $\langle\Theta\rangle$ is an infinite cyclic subgroup of $G_{GF(q)}(\overline{GF(q)})$.

(iv) There is a natural exact sequence of groups

$$1 \to G_{GF(q^s)}(GF(q^{st})) \to G_{GF(q)}(GF(q^{st})) \to G_{GF(q)}(GF(q^s)) \to 1$$

and this sequence corresponds to

$$1 \to \left\langle\Theta^s\right\rangle / \left\langle\Theta^{st}\right\rangle \to \langle\Theta\rangle / \left\langle\Theta^{st}\right\rangle \to \langle\Theta\rangle / \left\langle\Theta^s\right\rangle \to 1 .$$

("Exactness" of this sequence of groups is equivalent to

$$[G_{GF(q)}(GF(q^{st}))] / [G_{GF(q^s)}(GF(q^{st}))] \cong G_{GF(q)}(GF(q^s)) .$$

The mappings are natural: First the injection and then the map which takes Θ onto its restriction to $GF(q^s)$.)

(b) Let $F(q) = \{k \mid GF(q) \subseteq k \subseteq \overline{GF(q)}$ and k is a finite field$\}$.

Then, for the natural numbers N the map

$N \to F(q)$ defined by

$n \to GF(q^n)$

is a bijection satisfying:

m divides n if and only if $GF(q^m) \subseteq GF(q^n)$.

The proof of the above is straightforward and left to the exercises.

An irreducible polynomial f in $k[X]$ is <u>separable</u> if f has $\deg(f)$ distinct zeros in the algebraic closure $\overline{k}$ of k . The field k is <u>perfect</u> if every irreducible polynomial in $k[X]$ is separable. It is well-known that a field of characteristic p is perfect if and only if $k^p = \{a^p \mid a \in k\}$ equals k . Thus finite fields are perfect.

Section (B). Finite Division Rings.

This section provides a proof that every finite division ring is a field. Initially we must consider polynomials of the form $X^n - 1$ over the rationals Q .

It is easy to see that $X^n - 1$ has n distinct zeros in the complex field C and the zeros form a cyclic group T_n. If ϕ denotes the Euler ϕ-function then T_n has $\phi(n)$ generators. The generators of T_n are called the <u>primitive</u> <u>n-th</u> roots of unity. Let ζ denote a fixed primitive n-th root of unity. Then

$$T_n = \{\zeta^1, \zeta^2, \ldots, \zeta^{n-1}, \zeta^n = 1\}$$

Let

$$\Phi_n = \prod_{(k,n)=1} (X - \zeta^k)$$

in C[X]. The zeros of Φ_n are precisely the primitive n-th roots of unity. The polynomial Φ_n is called the <u>n-th</u> <u>cyclotomic</u> <u>polynomial</u>.

It is easy to show $X^n - 1 = \prod_{d|n} \Phi_d$.

<u>(II.7)</u> <u>LEMMA</u>. The polynomial Φ_n has rational integer coefficients.

<u>Proof</u>. The proof is by induction. Observe that Φ_n is monic and that $\Phi_1 = X - 1$ and $\Phi_2 = X + 1$.

Since $X^n - 1 = \prod_{d|n} \Phi_d$ we have $X^n - 1 = [\prod_{d|n, d \neq n} \Phi_d]\Phi_n$ and by induction

$f = \prod_{d|n, d \neq n} \Phi_d$ is in $Z[X]$.

Thus, since $Z[X]$ is a Euclidean domain, as polynomials in $Z[X]$

$$X^n - 1 = fg + r$$

with $r = 0$ or $\deg(r) < \deg(f)$. But $Z \subset Q[\zeta]$. Thus in $(Q[\zeta])[X]$, where $Q[\zeta]$ is the field generated by Q and ζ ,

$$X^n - 1 = fg + r$$

and thus $r = 0$.

(II.8) LEMMA. If $0 < r < n$ and $r|n$ then

$$\Phi_n \text{ divides } \left(\frac{X^n - 1}{X^r - 1}\right)$$

in $Z[X]$.

Proof. By (II.7) Φ_n is in $Z[X]$ and obviously Φ_n divides

$$X^n - 1 = (X^r - 1)\left(\frac{X^n - 1}{X^r - 1}\right).$$

But if $r|n$ then Φ_n and $X^r - 1$ have no common zeros. Hence Φ_n divides $\left(\frac{X^n - 1}{X^r - 1}\right)$.

A group G is said to act as a transformation group on itself by conjugation if for each g in G we define

$$\rho_g : G \to G$$

by

$$\rho_g : h \to ghg^{-1} = h^g .$$

For an element h in G let

$$[h] = \{h^g \mid g \in G\}$$

be the <u>orbit</u> of h and let

$$I(h) = \{g \in G \mid h^g = h\}$$

be the <u>stablizer</u> of h . Then $|[h]| = [G : I(h)]$ and it is easy to show since the orbits partition G that we have the <u>class equation</u>

$$|G| = \sum_h [G : I(h)]$$
$$= \left(\sum_{\substack{h \\ h \notin Z(G)}} [G : I(h)] \right) + |Z(G)| .$$

where $\sum_h$ extends over a complete set of elements from distinct orbits and Z(G) denotes the center of G .

<u>(II.9)</u> <u>THEOREM</u>. (Wedderburn) A finite division ring is a field.

<u>Proof</u>. Let D denote a finite division ring and let $k = \{z \mid zd = dz$ for all d in D$\}$ be the center of D . Clearly k is a finite field and hence for some $q = p^m$, $k = GF(q)$. The ring D is clearly a k-vector space of, say, dimension n . Thus $|D| = q^n$. If $n = 1$ then $D = k$ and we are done. Thus, suppose $n > 1$, and let the group of units

$D^* = D - \{0\}$ act on itself as a transformation group under conjugation. For a in D, $I(a) \cup \{0\}$ is a division ring and $k \subseteq I(a) \cup \{0\}$. Thus, for some r, $|I(a)| = q^r - 1$.

Since $I(a)$ is a subgroup of D^* and thus $q^r - 1$ divides $q^n - 1$ it is easy to show that r divides n.

By the class equation

$$|D^*| = \sum_a |[a]|$$

where

$$|[a]| = \begin{cases} (q^n - 1) / |I(a)| & a \notin k^* \\ 1 & a \in k^* \end{cases}.$$

Thus, since $|I(a)| = q^r - 1$ where r is a function of a,

$$q^n - 1 = (q - 1) + \sum \frac{q^n - 1}{q^r - 1}$$

where $\sum$ extends over certain divisors r of n with $0 < r < n$.

By (II.8) if $0 < r < n$ and $r|n$ we have Φ_n dividing $(X^n - 1) / (X^r - 1)$ and thus $\Phi_n(q)$ divides $(q^n - 1)/(q^r - 1)$. <u>Thus $\Phi_n(q)$ divides $q - 1$.</u>

But $\Phi_n = \prod_{(k,n)=1} (X - \zeta^k)$ and in C, $|\Phi_n(q)| = \prod_{(k,n)=1} |q - \zeta^k|$

$$\geq \Pi \big| |q| - |\zeta|^k \big|$$

$$= \Pi(q - 1) > q - 1$$

when $n > 1$ since $q = p^m \geq 2$.

Thus $\Phi_n(q)$ cannot divide $q - 1$ - a contradiction - and consequently we do not have $n > 1$.

Section (C). Skew-Polynomial Rings.

For this section we fix the following notation:

(1) k will denote a finite field and Θ will be an automorphism of k with $|\langle\Theta\rangle| = m$.

(2) k_0 will denote the subfield of k fixed under $\langle\Theta\rangle$. Thus, $m = [k : k_0]$. Further, if $k_0 = GF(q)$ where $q = p^t$ for some prime p , then $k = GF(q^m)$ and $\Theta : \alpha \to \alpha^q$ for all α in k .

For the above notation, let $k[X;\Theta]$ the set of all (skew) polynomials

$$\alpha_0 + \alpha_1 X + \cdots + \alpha_n X^n$$

where α_i are in k , X is an indeterminate and $n = 0,1,2,\ldots$. Equality and addition of these polynomials is defined in the standard manner while multiplication is given by the distributive law and

$$(\alpha X^i)(\beta X^j) = \alpha\Theta^i(\beta)X^{i+j} .$$

The set $k[X;\Theta]$ with the above operations forms a ring called the <u>skew-polynomial</u> ring over k with automorphism Θ .

If $f = \alpha_0 + \alpha_1 X + \cdots + \alpha_n X^n$ with $\alpha_n \neq 0$ is in $k[X;\Theta]$ then f is said to have <u>degree</u> n denoted $\deg(f) = n$. Let $\deg(0) = -\infty$. The next lemma is easy to verify. (With usual assumptions on symbol $-\infty$.)

<u>(II.10)</u> <u>LEMMA</u>. If f and g are in $k[X;\Theta]$, then

(a) $\deg(f + g) \leq \max\{\deg(f), \deg(g)\}$

(b) $\deg(fg) = \deg(f) + \deg(g)$.

The lemma implies that $k[X;\Theta]$ has no zero divisors and that the units of $k[X;\Theta]$ consist of the polynomials of degree 0 , i.e., the scalars k .

<u>(II.11)</u> <u>THEOREM</u>. (Division Algorithm) Let f and g be in $k[X;\Theta]$ with $f \neq 0$. Then there exist polynomials q and r with

$$g = qf + r$$

and

$$\deg(r) < \deg(f) .$$

<u>Proof</u>. If $\deg(g) < \deg(f)$, take $q = 0$ and $r = g$. Thus, assume $\deg(g) \geq \deg(f)$ and let

$$g = \alpha_0 + \alpha_1 X + \cdots + \alpha_t X^t ,$$
$$f = \beta_0 + \beta_1 X + \cdots + \beta_s X^s .$$

Then

$$g - \alpha_t \Theta^{t-s}(\beta_s^{-1})X^{t-s}f = \delta_0 + \delta_1 X + \cdots + \delta_{t-1}X^{t-1} .$$

The proof may now be completed by induction on the degree of the polynomial g .

The above result is "division on the right" by f . A similar argument will show that there exist polynomials q_1 and r_1 with

$$g = fq_1 + r_1 \quad \text{and} \quad \deg(r_1) < \deg(f) .$$

Select any non-zero left ideal I in $k[X;\Theta]$ and in I choose from among the non-zero polynomials one of least degree, say f . If g is any polynomial in I , $g = qf + r$ where $\deg(r) < \deg(f)$. But $r = g - qf$ is in I . Hence, by the minimality of $\deg(f)$, $r = 0$ and $g = qf$. Therefore I is a principal left ideal. Similarly every right ideal of $k[X;\Theta]$ is principal. Thus $k[X;\Theta]$ is a <u>non-commutative principal ideal domain</u>.

We now describe the two-sided ideals of $k[X;\Theta]$. Suppose I is a two-sided ideal. Then as a left ideal $I = k[X;\Theta]g$ and as a right ideal $I = fk[X;\Theta]$. Thus there exist polynomials s and t with $fs = g$ and $tg = f$. Since tf is in I , $tf = ft'$ for some t' . Then $f = tg = tfs = ft's$ and since $k[X;\Theta]$ is a domain, $1 = t's$ and s is a unit in k . Similarly t is a unit and $fk[X;\Theta] = k[X;\Theta]f = I$. That is, any left generator is a right generator and conversely.

Suppose $f = \alpha_0 X^t + \alpha_1 X^{t+1} + \cdots + \alpha_n X^{n+t}$ generates I . It is clear that X^t generates a two-sided ideal and it may be proven that $\alpha_0 + \alpha_1 X + \cdots + \alpha_n X^n$ then generates a two-sided ideal. Thus, we may assume that $f = \alpha_0 + \alpha_1 X + \cdots + \alpha_n X^n$ where $\alpha_0 \neq 0$. If β is in the field k then $\beta f = f\delta$ for some δ in $k[X;\Theta]$. By examining degrees, observe $\deg(\delta) = 0$ and δ is in k . Then

$$\begin{aligned} f\delta &= (\alpha_0 + \alpha_1 X + \cdots + \alpha_n X^n)\delta \\ &= \alpha_0\delta + \alpha_1\Theta(\delta)X + \cdots + \alpha_n\Theta^n(\delta)X^n . \end{aligned}$$

Since this is equal to βf , we must have

$$\beta = \delta, \beta = \Theta(\delta), \ldots, \beta = \Theta^n(\delta) .$$

But β was chosen arbitrarily in k , hence the order $|\langle\Theta\rangle| = m$ of Θ must divide each power of X in f , i.e.,

$$f = \alpha_0 + \alpha_1 X^m + \alpha_2 X^{2m} + \cdots + \alpha_n X^{nm} .$$

(II.12) THEOREM. If a polynomial f generates a two-sided ideal in $k[X;\Theta]$, then f has the form

$$(\alpha_0 + \alpha_1 X^m + \cdots + \alpha_n X^{nm})X^t$$

where $m = |\langle\Theta\rangle|$.

Observe $k_0[X] = \{\alpha_0 + \alpha_1 X + \cdots + \alpha_n X^n \mid n = 0,1,\ldots;\alpha_i \text{ in } k_0\}$ forms a commutative subring of $k[X;\Theta]$; and, if $k[X^m] = \{\alpha_0 + \alpha_1 X^m + \cdots + \alpha_n X^{mn} \mid n = 0,1,\ldots;\alpha_i \text{ in } k\}$ then the center of $k[X;\Theta]$ is $k_0[X] \cap k[X^m]$ and consists of all polynomials in X^m with coefficients in k_0 .

A second characterization of $k[X;\Theta]$ is provided by the Ore polynomial ring. Let

$$k[Y^q;o] = \{a_0Y + a_1Y^q + \cdots + a_nY^{q^n} \mid n = 0,1,\ldots;a_i \text{ in } k\} .$$

For

$$f = a_0Y + a_1Y^q + \cdots + a_nY^{q^n}$$

and

$$g = b_0Y + b_1Y^q + \cdots + b_tY^{q^n} ,$$

define $f + g$ to be ordinary addition of polynomials and define

$$f \circ g = f(g) = a_0g + a_1g^q + \cdots + a_ng^{q^n} .$$

Thus, if

$$f \circ g = c_0Y + \cdots + c_{n+t}Y^{q^{n+t}}$$

then

$$c_i = \sum_{j+s=i} a_jb_s^{q^j} .$$

It is easy to see that $k[Y^q;o]$ under addition and "composition" o forms a non-commutative ring, called the Ore polynomial ring.

Define

$$\Phi : k[X;\Theta] \to k[Y^q;o]$$

by

$$\Phi(\sum \alpha_i X^i) = \sum \alpha_i Y^{q^i} .$$

That is, $X \to Y^q, \ldots, X^i \to Y^{q^i}$ and note

$$(X^i)(\alpha X^j) = \Theta^i(\alpha) X^{i+j} = \alpha^{q^i} X^{i+j}$$

maps to $\alpha^{q^i} Y^{q^{i+j}} = (\alpha Y^{q^j})^{q^i} = Y^{q^i} \circ (\alpha Y^{q^i})$. The next theorem is now straightforward.

<u>(II.13)</u> <u>THEOREM</u>. The above mapping

$$\Phi : k[X;\Theta] \to k[Y^q;o]$$

determines a ring isomorphism between the skew-polynomial ring $k[X;\Theta]$ and the Ore polynomial ring $k[Y^q;o]$.

A polynomial f is called a <u>left multiple</u> of a polynomial d (and d a <u>right divisor</u> of f) if there exists a polynomial g with $f = gd$. Observe that if d is a right divisor of f in $k[Y^q;o]$, then $f = g \circ d = g(d) = \sum \alpha_i d^{q^i} = (\sum \alpha_i d^{q^i - 1})d$, i.e., d is a divisor of f in the ordinary sense. A polynomial d is a <u>greatest common right divisor</u> of f and g if (1) d is a right divisor of f and g and (2) d is a left multiple of every right divisor of f and g .

<u>(II.14)</u> <u>THEOREM</u>. The greatest common right divisor of two polynomials f and g in $k[Y^q;o]$ is equal to the ordinary greatest common divisor of f and g .

The above theorem follows immediately from the preceding remarks and serves to relate computation in $k[Y^q;o]$ (and hence in $k[X;\Theta]$) to the ordinary arithmetic of polynomials.

Section (D). Further Results on Finite Fields and Their Polynomial Rings.

We continue the setting of the previous section, that is,

(1) k denotes a finite field with automorphism Θ .

(2) k_0 denotes the fixed subfield of k under $\langle\Theta\rangle$ with $[k : k_0] = m = |\langle\Theta\rangle|$.

The finite field k is naturally a k_0-vector space of dimension $m = [k : k_0] = |\langle\Theta\rangle|$. Let $\mathrm{End}_{k_0}(k)$ denote the k_0-linear maps of k as a k_0-space ($\mathrm{End}_{k_0}(k)$ is isomorphic to $M_m(k_0)$). Observe

(1) There is a natural identification of k_0 with the center of $\mathrm{End}_{k_0}(k)$ by viewing α in k_0 as a right multiplication map.

(2) Similarly k may be identified with a subring of $\mathrm{End}_{k_0}(k)$ via right multiplication.

(3) The automorphism Θ is a k_0-linear map of k and $\langle\Theta\rangle$ is a subgroup of $GL_{k_0}(k)$ the invertible k_0-linear maps of k .

Our concern is with the subring generated by k and $\langle \Theta \rangle$ in $\mathrm{End}_{k_0}(k)$. Let (k,Θ) denote the subring generated by k and Θ in $\mathrm{End}_{k_0}(k)$. Note (k,Θ) is a k_0-algebra. Indeed, (k,Θ) is a <u>semi-linear group ring</u>, that is,

(1) (k,Θ) is a k-vector space with basis $\{1,\Theta,\Theta^2,\ldots,\Theta^{m-1}\}$.

(2) Multiplication in (k,Θ) is given by

$$\Theta^i \cdot \alpha = \Theta^i(\alpha)\Theta^i \qquad (\alpha \text{ in } k),$$
$$\Theta^i \cdot \Theta^j = \Theta^{i+j}$$

and the distributive property.

<u>(II.15)</u> <u>THEOREM</u>. $\mathrm{End}_{k_0}(k) = (k,\Theta)$. (Hence $(k,\Theta) \simeq M_m(k_0)$.)

<u>Proof</u>. Certainly (k,Θ) is a k_0-subalgebra of $\mathrm{End}_{k_0}(k)$. A k_0-basis of k together with $\{1,\Theta,\ldots,\Theta^{m-1}\}$ forms a k_0-basis of (k,Θ). Thus $\dim_{k_0}(k,\Theta) = [k : k_0]m = m^2 = \dim_{k_0}(\mathrm{End}_{k_0}(k))$. Hence $|(k,\Theta)| = |\mathrm{End}_{k_0}(k)|$ and $(k,\Theta) = \mathrm{End}_{k_0}(k)$.

We now have one of several applications of skew-polynomial rings.

<u>(II.16)</u> <u>COROLLARY</u>.

(a) $\mathrm{End}_{k_0}(k) \simeq k[X;\Theta]/(X^m - 1)$

(b) $\mathrm{End}_{k_0}(k) \simeq k[Y^q;o]/(X^{q^m} - X)$

where $k[X;\Theta]$ is the skew-polynomial ring and $k[Y^q;o]$ is the Ore polynomial ring of the preceding section.

Proof. Observe there is a natural ring morphism

$$\Pi : k[X;\Theta] \to (k,\Theta)$$

where

$$\Pi : \sum a_i X^i \to \sum a_i \Theta^i .$$

The kernel of Π is the two-sided ideal generated by $X^m - 1$.

Theorem (II.15) shows $(k,\Theta) = \mathrm{End}_{k_0}(k)$. It is natural to ask about the subfields of k containing k_0 and the subrings they in turn generate in $\mathrm{End}_{k_0}(k)$ together with Θ. In particular, what is (k_0,Θ) ? The structure of (k_0,Θ) was determined by Ore [95] in 1934; however, only recently Brawley, Carlitz and Vaughn [78] completed the problem.

Let k_1 be a subfield of k with $k_0 \subseteq k_1 \subseteq k$. Then there is an integer s with $m = ts$, $[k : k_1] = t$, $[k_1 : k_0] = s$, $k = GF(q^m) = GF(q^{ts})$, and $k_1 = GF(q^s)$ where $k_0 = GF(q)$. Further $(k_0,\Theta) \subseteq (k_1,\Theta) \subseteq (k,\Theta) = \mathrm{End}_{k_0}(k)$.

(II.17) THEOREM. (For the above setting.)

(a) (Brawley, Carlitz, Vaughn [78])

(k_1,Θ) is isomorphic to the $s \times s$ matrix ring $M_s(R)$ over $R = k_0[X]/(X^t - 1)$.

(b) (Ore [95]) In particular,

(k_0,Θ) is isomorphic to $k_0[X]/(X^m - 1)$.

Proof. If g is in (k_1,Θ) then

$$g = \sum_{v=0}^{m-1} a_v\Theta^v \qquad (a_v \text{ in } k_1)$$

$$= \sum_{i=0}^{t-1} \sum_{j=0}^{s-1} a_{is+j}\Theta^{is+j}$$

$$= \sum_{i=0}^{t-1} g_i(\Theta)\Theta^{is} = \sum_{i=0}^{t-1} \Theta^{is}\, g_i(\Theta)$$

where $g_i(X) = \sum_{j=0}^{s-1} a_{is+j}X^j$. Note $g_i(\Theta)$ is a k_0-linear map of k_1 .

Let G be in $M_s(R)$ where $R = k_0[X]/(X^t - 1)$. Then

$$G = G_0 + G_1X + \cdots + G_{t-1}X^{t-1}$$

where G_i is in $M_s(k_0)$.

Fixing a basis for k_1 over k_0 and regarding the G_i as k_0-linear maps of k_1 , it is easy to see each G_i corresponds to a $g_i = \sum_{j=0}^{s-1} b_{ij}\Theta^j$ in $\mathrm{End}_{k_0}(k)$.

Define

$$\Gamma : M_s(R) \to (k_1,\Theta)$$

by

$$\Gamma : G_iX^i \to g_i(\Theta)\Theta^{is}$$

and extending Γ linearly. It is now straightforward to check that Γ is a ring isomorphism.

To complete this section we examine Θ (the k_0-automorphism of k generating $G_{k_0}(k)$) as a k_0-linear mapping of k. As a linear mapping we would like to determine the centralizer $C(\Theta) = \{ \beta \mid \beta\Theta = \Theta\beta , \beta \text{ in } \mathrm{End}_{k_0}(k)\}$ of Θ in $\mathrm{End}_{k_0}(k)$ and the invariant subspaces of Θ. An acquaintance with elementary linear algebra is assumed for the remainder of (D).

The material we now prove was essentially described by Ore [94], [95] utilizing Ore polynomial rings. We select an equivalent approach via linear algebra techniques. This approach was recently used by Vaughn [101] to develop a variety of results.

(II.18) THEOREM. Let $C(\Theta)$ be the centralizer of Θ in $\mathrm{End}_{k_0}(k)$. Then

(a) $C(\Theta) = \{a_0 + a_1\Theta + \cdots + a_{m-1}\Theta^{m-1} \mid a_i \text{ in } k_0\}$

(b) $C(\Theta) \simeq k_0[X]/(X^m - 1)$.

Proof. Let β in $\mathrm{End}_{k_0}(k)$ satisfy $\beta\Theta = \Theta\beta$. Since $(k,\Theta) = \mathrm{End}_{k_0}(k)$, $\beta = a_0 + a_1\Theta + \cdots + a_{m-1}\Theta^{m-1}$ with $a_i \in k$. Requiring $\beta\Theta = \Theta\beta$, implies by direct computation that the a_i lie in k_0 . Conversely, any element $a_0 + a_1\Theta + \cdots + a_{m-1}\Theta^{m-1}$ with a_i in k_0 lies in $C(\Theta)$. Part (b) is immediate.

Let f be a monic polynomial in $k_0[X]$. The integer $\Phi(f)$ is defined to be the number of polynomials in $k_0[X]$ of degree less than the degree of f and relatively prime to f. Recall $|k_0| = q$. If the degree of f is t, define $|f| = q^t$. The following result is left as an exercise and should be compared with the Euler ϕ-function.

(II.19) THEOREM. (Dedekind) Suppose the monic polynomial f in $k_0[X]$ has the factorization $f = p_1^{e_1} \cdots p_s^{e_s}$ where the p_i are monic irreducible coprime polynomials in $k_0[X]$. Then

$$\Phi(f) = |f| \prod_{i=1}^{s} \left(1 - \frac{1}{|P_i|}\right).$$

Let a be in k. If the set $\{a, \Theta(a), \ldots, \Theta^{m-1}(a)\}$ forms a k_0-basis of k, then $\{a, \Theta(a), \ldots, \Theta^{m-1}(a)\}$ is called a normal basis for k over k_0. From (II.18) and (II.19) the number of normal bases for k over k_0 may be determined.

(II.20) THEOREM. Let N denote the number of distinct normal bases for k over k_0. Then

$$N = \frac{1}{m} \Phi(X^m - 1).$$

Proof. Let $\{a, \Theta(a), \ldots, \Theta^{m-1}(a)\}$ and $\{b, \Theta(b), \ldots, \Theta^{m-1}(b)\}$ be two normal bases of k over k_0. Define β in $\mathrm{End}_{k_0}(k)$ by $\beta(\Theta^i(b)) = \Theta^i(a)$ for $0 \leq i \leq m-1$. Then $\beta\Theta = \Theta\beta$ in $\mathrm{End}_{k_0}(k)$, i.e., β is in $C(\Theta)$, and thus β is a polynomial in Θ . Further, by (II.18), $f(\Theta)$ is invertible if and only if $(f(X),X^m - 1) = 1$. It follows that the number of distinct normal bases is $\frac{1}{m}\Phi(X^m - 1)$.

The action of the linear mapping Θ on the k_0-space k converts k into a $k_0[X]$-module[1] by

$$X \cdot a = \Theta(a) ,$$

more generally,

$$(a_0 + a_1 X + \cdots + a_n X^n) \cdot a = (a_0 + a_1\Theta + \cdots + a_n\Theta^n)(a)$$

for a in k . Since k has a normal basis over k_0 , it is easy to see k is a cyclic torsion $k_0[X]$-module with annihilator the ideal generated by $X^m - 1$.

Let a be in k with $\{a,\Theta(a),\ldots,\Theta^{m-1}(a)\}$ a normal basis of k over k_0 . Define

$$\Lambda : k_0[X]/(X^m - 1) \to k$$

by

$$\Lambda : f \to f(\Theta)(a) .$$

---

[1]For those unfamiliar with this term, a "module" over a ring and accompanying ideas are defined in (III.).

Then Λ is an isomorphism as $k_0[X]$-modules (and as k_0-spaces).

Let $\Pi : k_0[X] \to k_0[X]/(X^m - 1)$ be the natural projection onto the quotient ring. By the Correspondence Theorem there exists a natural bijection between the ideals of $k_0[X]/(X^m - 1)$ and the ideals of $k_0[X]$ containing $(X^m - 1)$, that is, I is an ideal of $k_0[X]/(X^m - 1)$ if and only if $I = \Pi(J)$ for J an ideal of $k_0[X]$ with $(X^m - 1) \subseteq J \subseteq k_0[X]$. Further the ideals of $k_0[X]/(X^m - 1)$ are precisely the $k_0[X]$-submodules of $k_0[X]/(X^m - 1)$.

Finally, a k_0-subspace V of k is <u>Θ-invariant</u> or <u>Θ-stable</u> if $\Theta(V) \subseteq V$, i.e., $\Theta(a)$ is in V for all a in V. The Θ-invariant subspaces of k are easily seen to be the $k_0[X]$-submodules of k.

Thus, by the above discussion, the Θ-invariant k_0-subspaces of k may be naturally identified with the ideals J in $k_0[X]$ with $(X^m - 1) \subseteq J \subseteq k_0[X]$. Since $k_0[X]$ is a Euclidean domain, J is an ideal with $(X^m - 1) \subseteq J \subseteq k_0[X]$ if and only if $J = (f)$ where f divides $X^m - 1$.

<u>(II.21)</u> <u>THEOREM</u>. The number of Θ-invariant k_0-subspaces of k is equal to the number of monic divisors of $X^m - 1$ in $k_0[X]$.

We conclude this section with some remarks about Θ-invariant subspaces of k.

(II.22) THEOREM. Let V be a Θ-invariant k_0-subspace of k and let $\{a,\Theta(a),\ldots,\Theta^{m-1}(a)\}$ be a normal basis of k over k_0. Then

(a) There exists a monic divisor f of $X^m - 1$ in $k_0[X]$ satisfying

$$V = \{[h(\Theta)f(\Theta)](a) \mid h \text{ in } k_0[X]\}$$
$$\simeq (f)/(X^m - 1) \qquad [\text{as a } k_0[X]\text{-module}].$$

Further, if $fg = X^m - 1$ and $\Theta|_V$ denotes the restriction of Θ to the subspace V, then

(b) g is the minimal and characteristic polynomial of $\Theta|_V$,

(c) $\dim(V) = \deg(g)$,

(d) $[f(\Theta)](a)$ is a cyclic vector for V, i.e., if $b = [f(\Theta)](a)$ then $\{b,\Theta(b),\ldots,\Theta^{s-1}(b)\}$ is a k_0-basis of V where $s = \deg(g)$,

(e) $V = \ker(g(\Theta))$.

Proof. As discussed above, V as a subspace of k may be identified (through the action of Θ on a) with the $k_0[X]$-submodule $(f)/(X^m - 1)$ of $k_0[X]/(X^m - 1)$.

The restriction $\Theta|_V$ of Θ to V is in turn described as multiplication by X on $(f)/(X^m - 1)$. That is, we have a natural $k_0[X]$-module morphism

$$\sigma : k_0[X]/(X^m - 1) \to (f)/(X^m - 1)$$

given by

$$\sigma : X \to X \cdot \bar{f}$$

where $\bar{f}$ denotes the coset $f + (X^m - 1)$.

This "multiplication by X " gives the following exact sequence of $k_0[X]$-modules.

$$0 \to (g)/(X^m - 1) \to k_0[X]/(X^m - 1) \xrightarrow{\sigma} (f)/(X^m - 1) \to 0 .$$

This sequence is exact as k_0-spaces, thus, for example,

$$\begin{aligned} \dim_{k_0}(V) &= \dim_{k_0}((f)/(X^m - 1)) \\ &= \dim_{k_0}[(k_0[X]/(X^m - 1))/((g)/(X^m - 1))] \\ &= \dim_{k_0}(k_0[X]/(g)) \\ &= \deg(g) . \end{aligned}$$

The theorem now follows easily.

(II.1) EXERCISE. Let $k = GF(q)$. For a given integer $m > 0$ show that

$$\sum_{x \in k} x^m = \begin{cases} -1 & \text{if } (q - 1) \mid m \\ 0 & \text{otherwise} \end{cases} .$$

(II.2) EXERCISE. Let $k = GF(q)$ and $k^* = k - \{0\}$. Let K be any field and $K^* = K - \{0\}$. Let $\sigma : k^* \to K^*$ be a group morphism (non-trivial). Show

$$\sum_{x \in k^*} \sigma(x) = 0$$

(II.3) EXERCISE. (Chevalley-Warning) Let $k = GF(q)$ and f be a polynomial of degree d in $k[X_1, \ldots, X_n]$. Let $N(f)$ denote the number of zeros of f in k. If $n > d$ show that $N(f) \equiv 0 \pmod p$. (Hint: (Ax) for each n-tuple x in $k^{(n)}$, show

$$1 + f(x)^{q-1} = \begin{cases} 1 & \text{if } f(x) = 0 \\ 0 & \text{otherwise} \end{cases}.$$

Summing over all x, show $N(f) \equiv -\sum f(x)^{q-1} \pmod p$. Now show $-\sum f(x)^{q-1} \equiv 0 \pmod p$ if $d < n$.) If f has no constant term and $n > d$ show that $N(f) \geq p$.

(II.4) EXERCISE. Prove Theorems (II.2) and (II.4).

(II.5) EXERCISE. Show that every $GF(q)$-algebra automorphism $\sigma : GF(q)[X] \to GF(q)[X]$ is of the form $\sigma : X \to a + bX$ where $b \neq 0$.

(II.6) EXERCISE. Show that every $GF(q)$-automorphism σ of the field of rational functions of $GF(q)[X]$ is of the form

$$\sigma : X \to \frac{a + bX}{c + dX}$$

where $ad - bc \neq 0$.

(II.7) EXERCISE. Prove Theorem (II.6).

(II.8) EXERCISE. Let $k = GF(q)$. Let N denote the number of solutions in $k^{(n)}$ of

$$a_0X_0^{n_0} + \cdots + a_rX_r^{n_r} = 0 .$$

For each i , let $d_i = \text{g.c.d.}(n_i, q - 1)$; for each i and each u in k , let $N_i(u)$ denote the number of solutions of $X^{n_i} = u$. Show $N_i(u)$ is 1 for $u = 0$ and otherwise is equal to d_i or 0 according to whether u is or is not a d_i-th power in k . Set $L(u) = \sum_{i=0}^{r} a_i u_i$ and show

$$N = \sum_{L(u)=0} N_0(u_0) \cdots N_r(u_r)$$

where the sum extends over all $u = (u_0, \ldots, u_r)$ with $L(u) = 0$.

(II.9) EXERCISE.

(a) Show that for a in Z/Zp^n $(n > 1)$ the mapping given by

$$\sigma(a) = \begin{cases} 0 & a = 0 \\ 1 & a \neq 0 \end{cases}$$

cannot be represented as a polynomial (see (II.4)). (Hint: For any polynomial f in $Z/Zp^n[X]$, $f(a + p) \equiv f(a) \pmod{p}$.)

(b) (Carlitz) Show that a map $\sigma : Z/Zp^n \to Z/Zp^n$ can be represented by a polynomial in $(Z/Zp^n)[X]$ if and only if

$$\sum_{s=0}^{r} (-1)^{r-s} \binom{r}{s} \sigma(c + s) \equiv 0 \mod p^{\nu(r)}$$

for all c in Z/Zp^n and all $r \geq 0$ where $\nu(r) = \min(n,\alpha(r))$ and $\alpha(r)$ is the highest power of p dividing $r!$.

(II.10) EXERCISE. (Carlitz) Let $k = GF(q)$, $q = p^n$. Let $f = X^m + a_1 X^{m-1} + \cdots + a_m$ be monic and define $|f| = p^{nm}$. Note the number of monic polynomials of degree m is p^{nm} .

(a) Let

$$\zeta(s) = \sum_{f \text{ monic}} \frac{1}{|f|^s} = \sum_{m=0}^{\infty} \frac{p^{nm}}{p^{nms}} .$$

Show that $\zeta(s)$ converges absolutely for $s > 1$ to $1/[1 - 1/p^{n(s-1)}]$.

(b) Show that $\zeta(s) = \Pi(1/(1 - 1/|f|^s))$ and that this product converges absolutely for $s > 1$.

(c) Let $\nu(m)$ denote the number of monic irreducible polynomials of degree m . Show that from (a) and (b),

$$1 - 1/p^{n(s-1)} = \prod_{f \text{ monic}} (1 - 1/|f|^s)$$

$$= \prod_{m=1}^{\infty} (1 - 1/p^{nms})^{\nu(m)} .$$

Take logarithms and show that

$$\sum_{m,t=1}^{\infty} \nu(m)/tp^{mtns} = \sum_{m=1}^{\infty} 1/mp^{nm(s-1)}$$

and thus

$$\sum_{d|m} d\nu(d) = p^{nm} .$$

Show the last is equivalent by Möbius Inversion to (II.3)(b).

(II.11) EXERCISE. (Möbius Inversion) Let P denote the positive integers and C denote the complex field.

The map $\mu : P \to C$ given by

$$\mu(n) = \begin{cases} 1 & \text{if } n = 1 \\ (-1)^t & \text{if } n \text{ is a product of } t \text{ distinct primes} \\ 0 & \text{if } n \text{ has a repeated prime factor} \end{cases} .$$

(a) Show $\sum_{d|n} \mu(d) = \begin{cases} 1 & n = 1 \\ 0 & n > 1 \end{cases}$ (where $\sum_{d|n}$ means summation is over all positive divisors of n).

(b) Show $\sum_{d|n|m} \mu\left(\frac{m}{n}\right) = \begin{cases} 1 & \text{if } d = m \\ 0 & \text{if } d \text{ is a proper divisor of } m \end{cases}$.

(c) [Additive Inversion] Let $f : P \to C$ be a mapping and suppose $g : P \to C$ is given by

$$g(n) = \sum_{d|n} f(d) .$$

Show $f(n) = \sum_{d|n} \mu(\frac{n}{d})g(d)$.

(d) [Multiplicative Inversion] Let $f : P \to C$ be a mapping and suppose $g : P \to C$ is given by

$$g(n) = \prod_{d|n} f(d) .$$

Show

$$f(n) = \prod_{d|n} g(d)^{\mu\left(\frac{n}{d}\right)} .$$

<u>(II.12)</u> <u>EXERCISE</u>. (Stewart) For this exercise one needs a knowledge of row equivalence of $m \times m$ matrices, the row-reduced echelon form (for example, see Cullen [9], pp. 25-29), and the row rank of a matrix. Let $P(m)$ denote the number of row-reduced canonical forms in $M_m(k)$ ($k = GF(q)$, $q = p^n$) and let $P(r,m)$ denote the number of canonical forms counted by $P(m)$ of row rank r . Show the following:

(a) $P(m) = \sum_{r=0}^{m} P(r,m)$.

(b) $P(0,m) = P(m,m) = 1$, $m = 1,2,3,\ldots$; $P(1) = 2$, $P(2) = 3 + q$.

(c) $P(r + 1,m + 1) = P(r,m) + q^{r+1}P(r + 1,m)$, $0 \leq r < m$.

(d) Let $S(r,m) = (q^{m+1-r} - 1)/(q^r - 1)$, $0 < r \leq m$, and $S(0,m) = 1$. Then $P(r,m) = S(r,m)P(r - 1,m)$, $0 < r \leq m$. Thus $P(r,m) = S(0,m)\, S(1,m) \cdots S(r,m)$.

(e) $P(m + 1) = 2P(m) + (q^m - 1)P(m - 1)$, $m \geq 2$.

(f) Let $T(r,m) = (q^m - 1)/(q^r - 1)$, $0 < r \leq m$, and $T(0,m) = 1$. Then $P(r,m) = T(r,m)P(r - 1,m - 1)$.

(g) Let $R(r,m)$ denote the number of matrices row equivalent to a canonical matrix of rank r . Then

$$R(r,m) = \frac{|GL_m(k)|}{q^{r(m-r)}|GL_{m-r}(k)|}$$

and

$$q^{m^2} = \sum_{r=0}^{m} R(r,m)P(r,m) \qquad \text{(Class Equation)} .$$

We remark on the relation of the above to finite projective geometry. Any non-zero row of a matrix A in $M_m(k)$ can be considered a point in an $m - 1$ dimensional projective space P_{m-1} over k . The rows of a matrix of rank r serve as a basis for an $r - 1$ dimensional subspace. Row-equivalent matrices represent the same subspace and conversely. Thus when $r \geq 1$, $m \geq 2$ there are $P(r,m)$ subspaces of dimension $r - 1$ in the projective space P_{m-1} over k ; each of these subspaces contains $S(1,r)$ points, and in each are $|GL_r(k)|/(q - 1)^r r!$ sets of r points that might serve as a basis for the subspace. Then $P(m)$ is the number of elements in the lattice of subspaces of P_{m-1} .

(II.13) EXERCISE. (Factorization of polynomials over a finite field) (Berlekamp) We know that every polynomial in $k[X]$ where $k = GF(q)$ factors into a product of powers of irreducible polynomials. However, for a given polynomial obtaining a factorization is no easy task. Below we give an algorithm due to Berlekamp [5].
Let $f = a_0 + a_1X + \cdots + a_mX^m$ be in $k[X]$ where $k = GF(q)$. Let

Q be an m by m matrix over k whose i-th row is given by the coefficients of $X^{q(i-1)}$ modulo (f) .

(a) Let $g = b_0 + b_1X + \cdots + b_{m-1}X^{m-1}$. Show the coefficients of g^q modulo (f) are given by $[b_0,b_1,\ldots,b_{n-1}]Q$.

(b) Show similarly the coefficients of $g^q - g$ modulo (f) are given by $[b_0,b_1,\ldots,b_{n-1}](Q - I_m)$.

(c) Show the coefficients of g lie in the null space of $Q - I_m$ if and only if $g^q - g \equiv 0$ modulo (f) .

(d) Assume the coefficients of g determine a vector in the null space of $Q - I_n$. Then f divides $g^q - g$.

(i) Show $g^q - g = \prod_{\alpha \text{ in } k} (g - \alpha)$. Thus f divides $\Pi(g - \alpha)$.

(ii) Deduce f divides $\prod_{\alpha \text{ in } k} (\text{g.c.d.}(f,g - \alpha))$.

(iii) If $\alpha \neq \beta$ show $g - \alpha$ and $g - \beta$ are coprime.

(iv) Deduce $\Pi(\text{g.c.d.}(f,g - \alpha))$ divides f . Thus

$$f = \prod_{\alpha \text{ in } k} (\text{g.c.d.}(f,g - \alpha)) .$$

By repreating the above process on each factor g.c.d.$(f,g - \alpha)$ we eventually obtain a factorization of f .

(e) Show the number of distinct irreducible factors of f is the dimension of the null space of $Q - I_m$.

(For further results on factorizations, the resultant and the discriminant see Berlekamp [5].)

(II.14) EXERCISE. Let V be an n-dimensional vector space over $k = GF(q)$.

(a) Show there is a natural k-vector space isomorphism between V and $k[X]/(X^n - 1)$ given by

$$\langle a_0, \ldots, a_{n-1} \rangle \to a_0 + a_1 X + \cdots + a_{n-1} X^{n-1} .$$

(b) A subspace W in V is cyclic if whenever $\langle a_0, a_1, \ldots, a_{n-1} \rangle$ is in W then $\langle a_{n-1}, a_0, a_1, \ldots, a_{n-2} \rangle$ is in W. Show cyclic subspaces correspond to ideals in $k[X]/(X^n - 1)$ and conversely.

(c) Determine the number of cyclic subspaces in V .

(II.15) EXERCISE. Let $C(a_1, \ldots, a_m)$ denote the cyclic $n \times n$ matrix whose first row is $[a_1, \ldots, a_m]$ for a_i in $k_0 = GF(q)$.

(a) Show that $C(a_1, \ldots, a_m)$ may be written as a polynomial in the matrix T where

$$T = \begin{bmatrix} 0 & 0 & 0 & \cdots & 0 & 1 \\ 1 & 0 & 0 & \cdots & 0 & 0 \\ 0 & 1 & 0 & \cdots & 0 & 0 \\ \vdots & \vdots & \vdots & & \vdots & \vdots \\ 0 & 0 & 0 & \cdots & 1 & 0 \end{bmatrix}$$

(b) Let σ generate the k_0-field automorphisms of $GF(q^m) = k$. Let $\{a, \sigma(a), \ldots, \sigma^{m-1}(a)\}$ be a normal basis of k over k_0 .

Show T is the matrix of σ (as a k_0-linear map of k) relative to the normal basis.

(c) Show the centralizer $C(\sigma)$ of σ in $\mathrm{End}_{k_0}(k)$ is isomorphic to the algebra of all cyclic matrices in $M_m(k_0)$.

(II.16) EXERCISE. (Dickson) Let $q = p^n$ and define $f : GF(q) \to GF(q)$ by $f(\alpha) = a_0 + a_1\alpha^q + \cdots + a_{t-1}\alpha^{q^{t-1}}$ where the a_i are in $GF(q)$. Show f is a bijection if and only if the determinant

$$\det \begin{bmatrix} a_0 & a_1 & \cdots & a_{t-1} \\ a_0^q & a_1^q & \cdots & a_{t-1}^q \\ \vdots & \vdots & & \vdots \\ a_0^{q^{t-1}} & a_1^{q^{t-1}} & & a_{t-1}^{q^{t-1}} \end{bmatrix} \neq 0 .$$

(II.17) EXERCISE. (Similarity in $k[X;\sigma]$) Let $S = k[X;\sigma]$ where $k[X;\sigma]$ is described in Section (C). Let f and g be in $k[X;\sigma]$ and suppose we have an S-isomorphism $\phi : S/Sf \to S/Sg$.

(a) Show $\deg(f) = \deg(g)$.

If $\phi : 1 + Sf \to u + Sg$ then for any h in S , $\phi(h(1 + Sf)) = hu + Sg$. In particular, since $0 + Sf \to 0 + Sg$, we have $f + Sf \to fu + Sg$ and fu is in Sg . Thus $fu = vg$ for some v in S . Let $m = fu = vg$.

The coset $1 + Sg$ has the form $wu + Sg$ for some w in S. Thus $wu = 1 + yg$. Hence, the greatest common right divisor of u and g is 1.

(b) Show the least common left multiple of u and g is m. Thus, (Ore) f and g are <u>left similar</u> if there exists a u in $S = k[X;\sigma]$ with $1 = (u,g)$ and $f = [u,g]/u$, equivalently $Sfu = Su \cap Sg$ and $Su + Sg = S$.

(c) Show that m is the least common right multiple of f and v and the greatest common left divisor of f and v is 1.

Thus f and g are <u>right similar</u>.

(d) Show for f and g in $k[X;\sigma]$ the following are equivalent:

(1) $S/Sf \simeq S/Sg$ (as left S-modules).

(2) $S/fS \simeq S/gS$ (as right S-modules).

(3) f and g are right similar.

(4) f and g are left similar.

Because of (d) f and g in $k[X;\sigma]$ are called simply <u>similar</u> if there is a u in $k[X;\sigma]$ with $Sfu = Su \cap Sg$ and $Su + Sg = S$.

(e) Determine when $X - \alpha$ and $X - \beta$ for α, β in $GF(q)$ are similar.

[For addition results on similar polynomials, factorization and left ideals see Ore [93],Jacobson [15], and Cohn [8].]

<u>(II.18)</u> <u>EXERCISE</u>. Prove Theorems (II.13) and (II.19).

(II.19) EXERCISE. (Algebraic Closure of Z/Zp) We have assumed an acquaintance with the concept of the algebraic closure of a field. However for finite fields this may be developed easily.

(a) Show that if $t < s$ (positive integers) then $t! \mid s!$

(b) By (a) we have a natural ascending chain of fields

$$Z/Zp = GF(p) \subset GF(p^{2!}) \subset GF(p^{3!}) \subset \cdots .$$

Set $GF(p^\infty) = \bigcup_n GF(p^{n!})$. Show

(1) $GF(p^\infty)$ is naturally a field.

(2) $GF(p^n)$ is a subfield of $GF(p^\infty)$ for every positive integer n .

(3) Every polynomial over $GF(p^\infty)$ of degree d has d zeros in $GF(p^\infty)$.

The field $GF(p^\infty)$ is called the algebraic closure of Z/Zp .

(II.20) EXERCISE. (Automorphisms of the algebraic closure of Z/Zp). From the previous exercise let $k_\infty = GF(p^\infty)$ and $k_n = GF(p^n)$, $n \geq 1$ and consider $G_{k_1}(k_\infty)$ the group of k_1-automorphisms of k_∞ . If Θ is the mapping $x \to x^p$ of k_∞ then Theorem (II.6) indicates that Θ generates an infinite cyclic subgroup of $G_{k_1}(k_\infty)$ and that the automorphism groups of the finite fields are induced by $\langle \Theta \rangle$. Let G_n denote $G_{k_1}(k_n)$ $(n \geq 1)$ and G_∞ denote $G_{k_1}(k_\infty)$.

(a) Partially order the positive integers N by $n \subset m$ if $n \mid m$, i.e., if n divides m . Show for every m and n

in N there is a t in N with $n \subset t$ and $m \subset t$.

(b) If $n \subset m$ then $k_n \subset k_m$. Show there is a natural group morphism

$$\Pi_{mn} : G_m \to G_n$$

where Π_{mm} = identity on G_m . [Hint: $\langle\Theta\rangle \to G_m \to G_n$ by $\Theta \to \Theta|_{k_m} \to (\Theta|_{k_m})|_{k_n} = \Theta|_{k_n}$. Equivalently, $\langle\Theta\rangle \to \langle\Theta\rangle/\langle\Theta^m\rangle \to (\langle\Theta\rangle/\langle\Theta^m\rangle)/(\langle\Theta^n\rangle/\langle\Theta^m\rangle)$ where the last group is isomorphic to $\langle\Theta\rangle/\langle\Theta^n\rangle$.]

(c) From (b) we have a family of groups and group morphisms $\{(G_n;\Pi_{mn}) \mid n \in N$ and $m \supset n\}$. Form the <u>direct product</u> ΠG_n of the G_n . This is the set of mappings

$$\{\sigma : N \to \bigcup_n G_n \mid \sigma(i) \text{ is in } G_i\}$$

and where $(\sigma\beta)(i) = \sigma(i)\beta(i)$. A map $\sigma : N \to \bigcup_n G_n$ is denoted as a "sequence" $\sigma = \langle\sigma_i\rangle_{i\in N}$ where $\sigma_i = \sigma(i)$. Inside ΠG_n take the set

$$G = \{\langle\sigma_i\rangle \mid \text{whenever } i \subset j \text{ then } \Pi_{ji}(\sigma_j) = \sigma_i\} .$$

Show G is a subgroup of ΠG_n . The subgroup G is called the <u>projective limit</u> of $\{(G_n;\Pi_{mn})\}$ and denoted $\operatorname{projlim}(G_n)$.

(d) Show $G_{k_1}(k_\infty) = G_\infty \simeq \operatorname{projlim}(G_n)$. [Hint: Let β be in G_∞ . Show $\beta|_{k_n}$ is in G_n . Show $\beta \to \beta|_{k_n}$ gives a surjective group morphism $G_\infty \to G_n$. Show this gives a natural group morphism

$\tau : G_\infty \to \Pi G_n$ by $\beta \to \langle \beta_i \rangle$ where $\beta_i = \beta|_{k_i}$. Show the image of τ is a subgroup of projlim(G_n) . Show τ is injective. Now let $\langle \sigma_i \rangle$ be in projlim(G_n) . Define $\sigma : k_\infty \to k_\infty$ by $\sigma(x) = \sigma_i(x)$ if x is in k_i . Show σ is a well-defined k_1-automorphism of k_∞ and $\tau(\sigma) = \langle \sigma_i \rangle$. Thus $G_\infty \simeq$ projlim(G_n) .]

<u>(II.21)</u> <u>EXERCISE</u>. Let $k[X;\sigma]$ be the skew-polynomial ring described in (C). If f is in $k[X;\sigma]$ and

$$f = q_1 \, q_2 \cdots q_t ,$$
$$f = p_1 \, p_2 \cdots p_s$$

where the p_i and q_j are irreducible, show $s = t$ and the p's and q's are similar in pairs (see Exercise (II.17)).

<u>(II.22)</u> <u>EXERCISE</u>. (Jacobson) Let C denote the center of $k[X;\sigma]$. Show that for every f in $k[X;\sigma]$, there exists a g in $k[X;\sigma]$ with fg in C . (g is called the <u>adjoint</u> of f and the existence of g implies that $k[X;\sigma]$ is a <u>Hamilitonian</u> domain. Hint: For f consider

$$1f = f1 , \quad X^m f = fX^m, \ldots , \quad X^{tm} f = fX^{tm}$$

where $m = |\langle \sigma \rangle|$. By the division algorithm

$$X^{tm} = fu_t + v_t .$$

The polynomials $v_0, v_1, \ldots, v_t, \ldots$ all have $\deg(v_i) < \deg(f)$. Thus, they cannot be k_0-linearly independent where k_0 is the fixed field of k under $\langle\sigma\rangle$. Therefore there exist $\alpha_0, \alpha_1, \ldots, \alpha_s$ in k_0 with $\sum \alpha_i v_i = 0$. Then $\sum \alpha_i X^{im} = f(\sum \alpha_i u_i) = fg$.)

(II.23) EXERCISE. (Jacobson) Let g be in the center of $k[X;\sigma]$. If $g = p_1 \cdots p_t$ where the p_i are irreducible, show the p_i are all similar (see Exercise (II.17)).

(II.24) EXERCISE. Let $S = k[X;\sigma]$ and f be an element of S . Clearly Sf is a left ideal but not in general an ideal or an algebra. To handle this one introduces the concept of an eigen-ring. Let R be a ring (not necessarily finite) and L a left ideal of R . The set $I(L) = \{r \text{ in } R \mid Lr \subseteq L\}$ is called the idealizer of L in R .

(a) Show I(L) is the largest subring of R in which L is a two-sided ideal.

The quotient ring $E(L) = I(L)/L$ is called the eigen-ring of L .

(b) (Cohn). Show $E(L) \simeq \mathrm{End}_R(R/L)$.

(c) (Cohn). Show the centralizer of r in R is isomorphic to the eigen-ring of $X - r$ in $R[X]$.

(d) If f is irreducible in S , show the eigen-ring of Sf is a finite field.

(II.25) EXERCISE. (Automorphisms of $k[X;\sigma]$) Let Λ be a k-ring automorphism of $k[X;\sigma]$. Then Λ is determined by its action on the element X . If

$$\Lambda : X \to f = s_0 + s_1X + \cdots + s_nX^n ,$$

denote Λ by Λ_f .

(a) Show $f = sX$ for some $s \neq 0$ in k .

(b) Show Λ_f is an automorphism if and only if $f = sX$ where s is a unit of k .

(Note: We are assuming $\sigma \neq$ identity .)

Suppose Φ is a ring automorphism of $k[X;\sigma]$. Then $\Phi(s_0 + s_1X + \cdots + s_nX^n) = \Phi(s_o) + \cdots + \Phi(s_n)\Phi(X)^n$. Let $\Phi(X) = f$ and define Λ_f by $\Lambda_f : X \to f$.

(c) Show Φ is an automorphism if and only if Λ_f is a k-ring automorphism.

(d) Show $\Phi|_k$ is an automorphism of k .

Thus Φ is composed of an automorphism of k and a k-ring automorphism of $k[X;\sigma]$.

III. FINITELY GENERATED MODULES OVER A RING

This section summarizes the basic terminology of finitely generated modules over a finite ring and examines Jordan chains, the Krull-Schmidt Theorem and projective modules.

For our purposes these facts will be utilized as follows: A finite ring R will be viewed as a finitely generated left R-module ${}_R R$ over itself. The module ${}_R R$ will decompose as a direct sum of left ideals. The uniqueness of such a decomposition is provided by the Krull-Schmidt Theorem. The ideals being summands of the free R-module ${}_R R$ are thus projective. Further, the Jordan-Hölder Theorem will provide information on the lattice of R-submodules of the individual left ideals.

Section (A). Preliminaries on Modules.

This subsection summarizes the basic terminology and elementary properties of a module over a ring. Probably the reader has encountered the concept of a ring module. Indeed, modules over polynomial rings were considered in (II). Due to this, our presentation will be terse. Additional detail may be obtained in most beginning algebra texts.

Let R be a ring and M an Abelian group written additively. Then M is called a left R-module if for each r in R and m in M there is a unique element rm in M satisfying

(a) $r(m + n) = rm + rn$

(b) $(rs)(m) = r(sm)$

(c) $(r + s)(m) = rm + sm$

(d) $1m = m$

for r and s in R and m and n in M.

A right R-module is defined similarly. For convenience we refer to left (or right) R-modules as R-modules. Often the prefix "R-" will be omitted.

Let M be an R-module. Let $N \subseteq M$. If N is an Abelian subgroup and whenever r is in R and n is in N then rn is in N, then N is an R-submodule of M. If N is a submodule of M then the quotient group M/N is an R-module where $r(m + N) = rm + N$ and is called the quotient module. If $\{N_1,\ldots,N_t\}$ is a collection of submodules of M, then $\cap N_i$ and $\sum N_i = N_1 + \cdots + N_t = \{n_1 + \cdots + n_t \mid n_i \text{ in } N_i\}$ are submodules of M. If $S \subseteq M$ then $\langle S\rangle$ denotes the smallest submodule of M containing S and is called the submodule generated by S. Observe

$$\langle S\rangle = \cap N$$

where N is a submodule of M with $S \subseteq N$. If $S = \{s_1,\ldots,s_n\}$ then $\langle S\rangle = \{r_1s_1 + \cdots + r_ns_n \mid r_i \text{ in } R\}$. If $\langle S\rangle = M$ then M is said to be generated by S. If $M = \langle S\rangle$ and $S = \{s_1,\ldots,s_n\}$, we write $M = Rs_1 + \cdots + Rs_n$ and say that M is finitely generated with generators $\{s_1,\ldots,s_n\}$. If $S = \{s\}$ and $M = Rs$ then M is called a cyclic R-module. If $M \neq 0$ has only two submodules, 0 and M, M is called a simple R-module. Note simple modules are cyclic.

NOTE: We will be concerned with finitely generated modules over finite rings.

If M and N are R-modules, then a mapping $\sigma : M \to N$ is called R-linear or an R-morphism if

(a) $\sigma(m_1 + m_2) = \sigma(m_1) + \sigma(m_2)$

(b) $\sigma(rm) = r\sigma(m)$

for r in R and m, m_1 and m_2 in M. If $\sigma : M \to N$ is an R-morphism, then

(a) the <u>kernel of σ</u>, $\ker(\sigma) = \{m \text{ in } M \mid \sigma(m) = 0\}$

(b) the <u>image of σ</u>, $\mathrm{Im}(\sigma) = \{\sigma(m) \mid m \text{ in } M\}$

are submodules of M and N, respectively. An R-morphism $\sigma : M \to N$ is called a <u>surjection</u> (<u>epimorphism</u>), an <u>injection</u> (<u>monomorphism</u>) or a <u>bijection</u> (<u>isomorphism</u>) according to $\mathrm{Im}(\sigma) = N$, $\ker(\sigma) = 0$ or both $\mathrm{Im}(\sigma) = N$ and $\ker(\sigma) = 0$, respectively. An isomorphism $M \to N$ is denoted by $M \simeq N$. If N is a submodule of M, there is a natural surjective R-morphism $M \to M/N$ given by $m \to m + N$.

If an R-morphism $\sigma : M \to M$ maps M to M then σ is called an <u>endomorphism</u> of M. If σ is bijective, σ is called an <u>automorphism</u> of M. If $\sigma^n = \sigma \cdot \sigma \cdots \sigma$ (n-factors) $= 0$ then σ is called <u>nilpotent</u>.

If $\sigma : M \to N$ is a surjective R-morphism then $N \simeq M/\ker(\sigma)$ and there is a natural bijective order preserving correspondence between the submodules of N and the submodules of M containing $\ker(\sigma)$.

If N and P are submodules of M then

(a) if $N \subseteq P \subseteq M$, then $M/N \simeq (M/P)/(N/P)$

(b) $(P + N)/P \simeq N/P \cap N$.

Let M be an R-module. Then $\mathrm{Annih}_R(M) = \{r \text{ in } R \mid rm = 0 \text{ for all } m \text{ in } M\}$ is a two-sided ideal in R called the <u>annihilator of M</u>. If m is in M, then $0(m) = \{r \text{ in } R \mid rm = 0\}$ is the <u>order left ideal</u> of m

If m is in M then m is a <u>torsion-free</u> element if $0(m) = 0$ and m is a <u>torsion</u> element if $0(m) \neq 0$. Further, M is <u>torsion</u> or <u>torsion-free</u> according to whether all its non-zero elements are <u>torsion</u> or <u>torsion-free</u>, respectively.

If M is an R-module with submodules $N_1, \ldots, N_t$ then M is called a <u>direct sum</u> (internal) of $N_1, \ldots, N_t$ if every element m of M can be expressed uniquely as

$$m = n_1 + \cdots + n_t$$

with n_i in N_i. Equivalently, $M = N_1 + \cdots + N_t$ and $N_i \cap (N_1 + \cdots + N_{i-1} + N_{i+1} + \cdots + N_t) = 0$ for $1 \leq i \leq t$.

If $N_1, \ldots, N_t$ are R-modules then the set of t-tuples $\langle n_1, \ldots, n_t \rangle$ with n_i in N_i is naturally an R-module with component-wise addition and $r\langle n_1, \ldots, n_t \rangle = \langle rn_1, \ldots, rn_t \rangle$. This module is called the <u>direct sum</u> (external) of $N_1, \ldots, N_t$.

If M is a direct sum of $N_1, \ldots, N_t$ then M is written as $M = \oplus \sum_{i=1}^{t} N_i = N_1 \oplus \cdots \oplus N_t$.

Let $\{M_i\}$ be a finite collection of R-modules and $\{\sigma_i\}$ a collection of R-morphisms with $\sigma_i : M_i \to M_{i+1}$. Then the sequence

$$\cdots \longrightarrow M_{i-1} \xrightarrow{\sigma_{i-1}} M_i \xrightarrow{\sigma_1} M_{i+1} \longrightarrow \cdots$$

is called exact if for each i, $\mathrm{Im}(\sigma_{i-1}) = \ker(\sigma_i)$. An exact sequence $M \xrightarrow{\sigma} N \longrightarrow 0$ $(0 \longrightarrow N \xrightarrow{\sigma} M)$ is said to split or splits if there exists an R-morphism $\beta : N \to M$ $(\beta : M \to N)$ with $\sigma\beta =$ identity on M ($\beta\sigma =$ identity on N) .

Section (B). Jordan-Hölder and Krull-Schmidt Theorems (Finitely Generated R-modules).

Recall that "ring" means finite ring. In this section we examine modules which are finitely generated over a ring R. Thus the module has a finite number of elements and trivially satisfies descending and ascending chain conditions on its submodules.

PROBLEM I: If M is a finitely generated R-module then we may naturally construct chains of submodules between the zero module and M, say

$$0 = M_0 \subset M_1 \subset \cdots \subset M_m = M$$

(with the M_i distinct) analogous to subspaces of a finite dimension vector space. A "maximal" chain could be obtained by inserting additional submodules between the existing ones until it is impossible to insert new submodules. Such chains obviously are finite in length. We are now concerned with the uniqueness of such maximal chains.

Let M be a finitely generated module over a ring R. A <u>normal chain</u> for M of <u>length</u> m is a chain

$$(*) \qquad 0 = M_0 \subset M_1 \subset \cdots \subset M_m = M$$

of submodules satisfying $M_i \neq M_{i+1}$ for $0 \leq i \leq m - 1$. A second normal chain

$$(**) \qquad 0 = N_0 \subset N_1 \subset \cdots \subset N_n = M$$

is a <u>refinement</u> of the first (*) if each M_i is some N_j, i.e., (**) is obtained from (*) by possibly inserting additional submodules. The quotient modules M_i/M_{i-1}, $1 \leq i \leq n$, of (*) are called the <u>composition factors</u> of the chain. Finally, (*) and (**) are <u>equivalent</u> if $m = n$ and there is a permutation σ of $\{0,1,\ldots,m-1\}$ such that

$$M_{i+1}/M_i \cong (N_{\sigma(i)+1})/\underline{N_i}$$

should be $N_{\sigma(i)}$?!

for $0 \leq i \leq m - 1$.

<u>(III.1)</u> <u>THEOREM</u>. (Schreier) Any two normal chains for a finitely generated R-module M have equivalent refinements.

<u>Proof</u>. This is easily proven using the Zassenhaus Lemma in the exercises.

A normal chain $0 = M_0 \subset M_1 \subset \cdots \subset M_m = M$ is a <u>composition chain</u> if it cannot be properly refined. The next result is an easy consequence of (III.1).

(III.2) THEOREM. (Jordan-Hölder) Let

$$0 = N_0 \subset N_1 \subset \cdots \subset N_n = M$$

$$0 = M_0 \subset M_1 \subset \cdots \subset M_m = M$$

be two composition chains for a finitely generated R-module M . Then $m = n$ and there is a permutation σ of $\{0,\ldots,m - 1\}$ such that

$$M_{i+1}/M_i \simeq N_{\sigma(i)+1}/N_i \qquad \text{(as R-modules)}$$

for $0 \leq i \leq m - 1$.

We now examine the decomposition of a finitely generated R-module into a direct sum of submodules. Let M be a finitely generated R-module. If $M = N \oplus P$ where N and P are non-zero submodules of M , then M is called decomposable. If M is not decomposable, it is called indecomposable.

PROBLEM II:

If M is a finitely generated R-module then it is indecomposable or it is not. If not, $M = N \oplus P$. Then N and P are each either indecomposable or decomposable. Continuing, we find

$$M = M_1 \oplus \cdots \oplus M_t$$

where M_i are indecomposable R-modules. We are now concerned with the uniqueness of such a decomposition. Observe

$|M| = |M_1| \cdots |M_t|$.

A decomposition of a module will be seen to be related to its R-morphisms. Thus we first examine R-linear maps of finitely generated R-modules.

(III.3) THEOREM. Let M be a finitely generated R-module and $\sigma : M \to M$ an R-morphism. Then the following are equivalent:

(a) σ is bijective,

(b) σ is injective,

(c) σ is surjective.

Proof. Trivial since $|M| < \infty$.

(III.4) THEOREM. (Fitting's Lemma) Let M be a finitely generated R-module and $\sigma : M \to M$ an R-morphism. Then there is an integer $n > 0$ and submodules N and P satisfying

(a) $M = N \oplus P$

(b) $N = \sigma^n(M)$

(c) $P = \sigma^{-n}(0)$.

Hence $\sigma = \sigma|_N \oplus \sigma|_P$ where $\sigma|_N$ is invertible while $\sigma|_P$ is nilpotent.

Proof. The chain $M \supset \sigma(M) \supset \sigma^2(M)$ stops at say n , i.e., $\sigma^n(M) = \sigma^{n+1}(M)$. Thus σ restricted to $\sigma^n(M)$ is surjective and hence by (III.3) an isomorphism. Set $N = \sigma^n(M)$ and $P = \sigma^{-n}(0)$. Clearly since $\sigma|_{\sigma^n(M)}$ is bijective, $N \cap P = 0$. To show $N + P = M$, select and m in M and choose m , with $\sigma^n(m) = \sigma^{2n}(m_1)$. Then

$m = \sigma^n(m_1) + (m - \sigma^n(m_1))$ is in $N + P$.

(III.5) COROLLARY. Let M be a finitely generated indecomposable R-module. Then any R-morphism of M is either nilpotent or an automorphism.

Let $\mathrm{End}_R(M)$ denote the set of R-morphisms $\sigma : M \to M$. Then under composition and addition of morphisms $\mathrm{End}_R(M)$ is a ring. Suppose that M is finitely generated and indecomposable. Let $\sigma_1, \ldots \sigma_n$ be in $\mathrm{End}_R(M)$ and suppose $\sigma_1 + \cdots + \sigma_n$ an automorphism. We claim that for some i , σ_i is an automorphism. Clearly it suffices to show this for $n = 2$. Thus, let $\beta = \sigma_1 + \sigma_2$ be an automorphism of M . Then $1 = \beta^{-1}\sigma_1 + \beta^{-1}\sigma_2$. If $\beta^{-1}\sigma_1$ is an automorphism, we are done. If not, by (III.5) $\beta^{-1}\sigma_1$ is nilpotent. Then $\beta^{-1}\sigma_2 = 1 - \beta^{-1}\sigma_1$ is an automorphism, since for a nilpotent δ , $1 - \delta$ is invertible with inverse $(1 - \delta)^{-1} = 1 + \delta + \delta^2 + \cdots$ (finite sum).

(III.6) THEOREM. (Krull-Schmidt) Let M be a finitely generated R-module. Suppose

$$M = M_1 \oplus \cdots \oplus M_m = N_1 \oplus \cdots \oplus N_n$$

are two decompositions of M as a direct sum of non-zero indecomposable submodules. Then $m = n$ are there exists a permutation σ of $\{1,\ldots,m\}$ such that $M_i \simeq N_{\sigma(i)}$ for $1 \leq i \leq m$.

Proof. Let $M = M_1 \oplus \cdots \oplus M_m = N_1 \oplus \cdots \oplus N_n$. For each $i = 1,\ldots,m$ and each $j = 1,\ldots,n$ there is a natural R-morphism $p_i : M \to M_i$ and $q_j : M \to N_j$ called the projection. For example, if x in M is written as $x = x_1 + \cdots + x_m$ with x_i in M_i then $p_i(x) = x_i$.

We now prove (III.6) by induction on the number of direct summands. Observe if x is in M then

$$x = \sum_{i=1}^{m} p_i(x) = \sum_{j=1}^{n} q_j(x) .$$

Thus the identity morphism on M is given by $i_M = \sum_{i=1}^{m} p_i$. Then $q_1 = \sum_{i=1}^{m} q_1 p_i$. Restricting this morphism to M_1 , we obtain $i_{M_1} = \sum_{i=1}^{m} q_1 p_i$. Since M_1 is indecomposable, by the discussion following (III.5) some $q_1 p_i$ is an automorphism of M_1 . Suppose without loss that $q_1 p_1$ is an automorphism. Then $N_1 \xrightarrow{p_1} M_1 \xrightarrow{q_1} N_1$ is an automorphism. By modifying (III.3) both p_1 and q_1 are isomorphisms. In particular, $|M_1| = |N_1|$.

It is easy to see $\overline{M} = N_1 + M_2 + \cdots + M_m$ is a direct sum. Clearly $\overline{M} \subseteq M$. But $|\overline{M}| = |N_1||M_2|\cdots|M_m| = |M_1||M_2|\cdots|M_m| = |M|$ since $N_1 \simeq M_1$. Thus $\overline{M} = M$. Define $\rho : M \to \overline{M} = M$ by $\rho = q_1 p_1 + p_2 + \cdots + p_m$. Then ρ is an automorphism with $\rho(M_1) = N_1$. Thus $M/M_1 \simeq \rho(M)/\rho(M_1) \simeq M/N_1$; that is, $M_2 \oplus \cdots \oplus M_m \simeq N_2 \oplus \cdots \oplus N_n$.

(III.7) COROLLARY. Let M be a finitely generated R-module and let $M = M_1 \oplus \cdots \oplus M_m$ be a decomposition of M into indecomposable non-zero submodules. If a submodule N of M is a direct summand of M then N is isomorphic to a direct sum of a subset of $\{M_1, \ldots, M_m\}$.

Additional discussion of finitely generated indecomposable modules is given in (XIII). The endomorphism ring $\mathrm{End}_R(M)$ where $M = M_1 \oplus \cdots \oplus M_m$ is examined in greater detail in (VII).

Section (C). Projective Modules.

A finitely generated R-module F is called R-free if there is a subset $\{b_i\}_{i=1}^n$ of F such that

(a) $Rb_i = \{rb_i \mid r \in R\} \simeq R$ (as left R-modules) under $b_i \to 1$.

(b) $F = \oplus \sum_i Rb_i$.

The set $\{b_i\}$ is called a free R-basis for F .

It is easy to see that if $B = \{b_i\}$ is a free basis for F and M is any R-module then any map $\alpha : B \to M$ determines naturally a unique R-morphism $\alpha^* : F \to M$ which extends α .

Further if F is free and we have the diagram of R-morphisms

$$\begin{array}{ccccc} & & F & & \\ & & \downarrow \beta & & \\ M & \xrightarrow{\alpha} & N & \longrightarrow & 0 \end{array}$$

where $M \xrightarrow{\alpha} N \longrightarrow 0$ is exact, i. e., $\mathrm{Im}(\alpha) = N$, then there is a morphism $\beta^* : F \to M$ with $\alpha \beta^* = \beta$.

Recall an exact sequence of left R-modules

$$M \xrightarrow{\alpha} N \longrightarrow 0$$

is said to split if there exists an R-morphism

$$\rho : N \to M$$

such that $\alpha\rho = i_N$ (i_N = identity map on N). If $M \xrightarrow{\alpha} N \longrightarrow 0$ (exact) splits with splitting morphism $\rho : N \to M$ then

(a) $N \simeq \mathrm{Im}(\rho)$ (as R-modules)

(b) $\mathrm{Im}(\rho) \cap \ker(\alpha) = 0$

(c) $M = \mathrm{Im}(\rho) + \ker(\alpha)$.

Thus $M = \mathrm{Im}(\rho) \oplus \ker(\alpha)$.

A finitely generated R-module P is called R-projective if whenever

$$\begin{array}{ccccc} & & P & & \\ & & \downarrow \beta & & \\ M & \longrightarrow & N & \longrightarrow & 0 \end{array} \qquad \text{(row exact)}$$

there exists an R-morphism $\beta^* : P \to M$ with $\alpha \beta^* = \beta$.

Clearly a free R-module is projective.

The following is straightforward.

(III.8) THEOREM. Let P be a finitely generated R-module. The following are equivalent.

(a) P is projective.

(b) Every exact sequence $M \to P \to 0$ splits.

(c) P is a direct summand of a free R-module.

We will employ (III.8) predominately in the following setting: Let R be a ring and L a left ideal of R. If L is a direct summand of ${}_R R$ (R regarded as a left R-module), i.e., ${}_R R = L \oplus L'$, then L is an R-projective module.

Projective modules over finite rings are characterized in (XIII).

(III.1) EXERCISE. (Zassenhaus Lemma) Let N, P, S, and T be submodules of a finitely generated R-module M. If $N \subset P$ and $S \subset T$, show

$$\frac{N + (P \cap T)}{N + (P \cap S)} \simeq \frac{S + (P \cap T)}{S + (P \cap N)} .$$

(III.2) EXERCISE. Let Z denote the rational integers. Determine the composition chains of the $Z/24Z$-module $Z/24Z$.

(III.3) EXERCISE. Let M be a finitely generated R-module and $0 = M_0 \subset M_1 \subset \cdots \subset M_m$ a composition chain for M. Then m is called the length of the module M and denoted $L_R(M)$.

(a) Let $P \subset N$ be submodules of M. Show

$$L_R(M/P) = L_R(M/N) + L_R(N/P) .$$

(b) Let N be a submodule of M. Show $L_R(N) \leq L_R(M)$; and, $L_R(M) = L_R(N)$ if and only if $M = N$.

(c) Let P and N be submodules of M. Show

$$L_R(P + N) + L_R(P \cap N) = L_R(P) + L_R(N)$$

and

$$L_R(M/(P + N)) + L_R(M/(N \cap P)) = L_R(M/P) + L_R(M/N) .$$

(III.4) EXERCISE.

(a) If $M \xrightarrow{\alpha} N \longrightarrow 0$ (exact) splits with splitting morphism β, show

$$M \simeq \mathrm{Im}(\beta) \oplus \ker(\alpha) \simeq N \oplus \ker(\alpha).$$

(b) If $0 \longrightarrow N \xrightarrow{\alpha} M$ (exact) splits with splitting morphism β, show

$$\begin{aligned} M &\simeq \mathrm{Im}(\alpha) \oplus \ker(\beta) \\ &\simeq N \oplus \ker(\beta) \\ &\simeq \mathrm{Im}(\beta) \oplus \ker(\beta) . \end{aligned}$$

(III.5) EXERCISE. Let M be a finitely generated R-module. Suppose $R = R_1 \oplus \cdots \oplus R_n$ (as rings). Show that M naturally decomposes as

$$M = M_1 \oplus \cdots \oplus M_n$$

where M_i is an R_i-module. Thus assume the characteristic of R is p^λ for a prime p. Show $\mathrm{End}_R(M) \subseteq \mathrm{End}_{(Z/Zp^\lambda)}(M)$. Show $\mathrm{End}_R(M)$ is a ring. If M and N are R-modules, let $\mathrm{Hom}_R(M,N)$ denote the set of R-morphisms $\sigma : M \to N$. Show $\mathrm{Hom}_R(M,N) \subseteq \mathrm{Hom}_{(Z/Zp^\lambda)}(M,N)$ and $\mathrm{Hom}_R(M,N)$ is a (Z/Zp^λ)-module.

(III.6) EXERCISE. Prove the following are equivalent for a finitely generated R-module P :

(a) P is R-projective.

(b) Every exact sequence

$$0 \longrightarrow L \xrightarrow{f} M \xrightarrow{g} N \longrightarrow 0 \qquad \text{(of modules)}$$

induces naturally an exact sequence

$$0 \longrightarrow \mathrm{Hom}_R(P,L) \xrightarrow{f^*} \mathrm{Hom}_R(P,M) \xrightarrow{g^*} \mathrm{Hom}_R(P,N) \longrightarrow 0 \qquad \text{(of groups)}$$

where, for example, $f^*(\sigma) = f\sigma$ for σ in $\mathrm{Hom}_R(P,L)$.

(c) Every exact sequence $M \to P \to 0$ splits.

(d) P is a direct summand of a free finitely generated R-module.

(e) There is a set $\{x_i\}_{i=1}^n$ in P and a set $\{\phi_i\}_{i=1}^n$ in $\mathrm{Hom}_R(P,R)$ such that

$$\sum_{i=1}^n \phi_i(p)x_i = p$$

for every p in P.

(III.7) EXERCISE. Show that every R-module is a quotient module of a projective R-module.

(III.8) EXERCISE. Let $M = \oplus \sum_{i=1}^{m} M_i$. Show that M is R-projective if and only if M_i is R-projective for every i , $1 \leq i \leq m$.

(III.9) EXERCISE. Let R be a ring and suppose that either

$$R = R_1 \oplus \cdots \oplus R_n \qquad \text{(as rings)}$$

or

$$R = L_1 \oplus \cdots \oplus L_t \qquad \text{(as left ideals)} .$$

In the first case, show that each R_i is R-projective (as either a left or right R-module) and, in the second case, show each L_i is a projective left R-module.

(III.10) EXERCISE. Let $k = GF(q)$ and let R be the ring of 2×2 matrices over k . Show that

$$\left\{ \begin{bmatrix} a & b \\ 0 & 0 \end{bmatrix} \;\middle|\; a,b \text{ in } k \right\} \text{ and } \left\{ \begin{bmatrix} a & 0 \\ b & 0 \end{bmatrix} \;\middle|\; a,b \text{ in } k \right\} \text{ are projective}$$

right and left R-modules, respectively.

(III.11) EXERCISE. Let $R = Z/nZ$. If r is a divisor of n we have a natural exact sequence

$$Z/Zn \xrightarrow{\alpha} rZ/Zn \longrightarrow 0$$

where the kernel of α is sZ/Zn and $s = n/r$. Show that as Z/Zn-modules this sequence splits if and only if g.c.d. $(r,s) = 1$.

<u>(III.12)</u> <u>EXERCISE</u>. Let F be a free R-module with basis $b_1,\ldots,b_n$ and let $\alpha = a_1b_1 + \cdots + a_nb_n$. Let A be the left ideal generated by $a_1,\ldots,a_n$. Show the following are equivalent.

(a) $A = Re$ for an idempotent e in R (e an idempotent means that $e^2 = e$).

(b) $R\alpha$, is an R-direct summand of F isomorphic to Re under $\alpha \to e$.

(c) There is an R-morphism $f : F \to Re$ with $f(\alpha) = e$ and $e\alpha = \alpha$.

IV. THE RADICAL

Several methods have evolved in a continuing search for the structure of rings.

If we seek a ring R which satisfies certain properties we might attempt to characterize the ideals or left ideals of the ring R and then show how these may be assembled to provide insight into the structure of R . Or, analogously, we may assume the modules of R satisfy certain properties; for example, require that all R-modules be free or be projective and deduce the structure of R .

For an arbitrary finite ring R we usually begin with a ring S which is assumed to be well-known; for example, a matrix ring or a polynomial ring. Then four approaches have emerged:

(1) We may attempt to view R as a ring homomorphic image of S . That is, seek an ideal N of S such that $R \simeq S/N$.

(2) We may attempt to embed R as a subring in S .

(3) We may attempt to break or cut from R a subring like S so that R = S + N where N is unknown but smaller and perhaps more easily handled.

(4) We may attempt to discover an ideal N in R such that $R/N \simeq S$.

Pictorially, we see structure as follows:

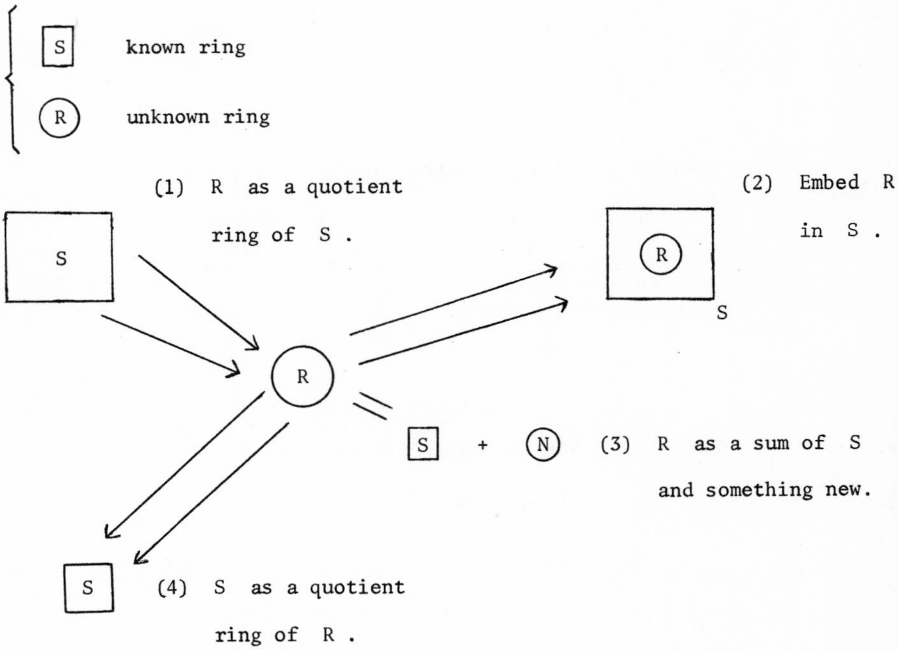

In this section, following the fourth approach, we develop an ideal N for a ring R such that the resulting quotient ring R/N is a direct sum of matrix rings over finite fields. Questions concerning R/N are questions of "linear algebra" and not generally questions of "ring theory". We then will attempt to "pull" or "lift" structure known for R/N to obtain a knowledge of R. A problem in ring theory is often considered solved if it may be reduced to a question of linear algebra. This ideal N of R will be called the radical of R.

Again, R denotes a finite associative ring with identity.

(IV.1) LEMMA. If a and b are in R and $ab = 1$ then $ba = 1$.

Proof. Consider the finite chain

$$bR \supseteq b^2R \supseteq \cdots \supseteq b^nR = b^{n+1}R = \cdots .$$

We note that $b^nR \neq 0$ since if $b^nR = 0$ then $b^na = 0$ and, by multiplying on the left by a^n, implies $a = 0$. Thus $b^nR = b^{n+1}R \neq 0$. Hence $b^{n+1}x = b^n$. Multiplying by a^n implies $bx = 1$. If $bx = 1$

then $x = a(bx) = a$, i.e., $ba = 1$.

We define the <u>radical of R</u> (Wedderburn, Artin), denoted Rad R or Rad(R) , to be the intersection of all maximal right ideals of R .

<u>(IV.2)</u> <u>PROPOSITION</u>.

$$\text{Rad } R = \{r \in R \mid 1 - rs \text{ is invertible for all } s \text{ in } R\} .$$

<u>Proof</u>. Let r be in Rad R . Then r is in M for every maximal right ideal M of R . Thus 1 is not in $M + rR$. Likewise, if 1 is not in $M + rR$ for every right maximal ideal M of R then r is in Rad R . Thus r in Rad R is equivalent to $1 - rs$ is in no maximal right ideal. Equivalently, $1 - rs$ is right invertible and hence, by (IV.1), invertible.

<u>(IV.3)</u> <u>LEMMA</u>. If $1 - rt$ is invertible then $1 - tr$ is invertible.

<u>Proof</u>. Let u be a unit with $(1 - rt)u = 1$. Then $(1 - tr)(1 + tur) = 1 + tur - t(1 + rtu)r = 1$.

The next proposition removes the seeming lack of symmetry of our definition.

<u>(IV.4)</u> <u>PROPOSITION</u>. The Rad R is an ideal and is the largest ideal K such that for all r in K , $1 - r$ is a unit.

Proof. The Rad R is clearly a right ideal. The above lemma implies that is a left ideal. The remainder is immediate.

(IV.5) PROPOSITION.

Rad R = $\cap$ M where M is a maximal left ideal of R .

(IV.6) PROPOSITION. Rad(R/Rad R) = 0 .

Proof. Let $\Pi : R \to R/\text{Rad } R$ be the natural morphism and let Πr be in Rad(R/Rad R) . Then r is in M for every maximal right ideal M of R containing the radical of R . Hence r is in Rad R and $\Pi r = 0$.

A ring R is called semi-simple if Rad R = 0 .

(IV.7) PROPOSITION. Rad R is the largest nilpotent ideal of R . Indeed, Rad R contains all nilpotent (left, right) ideals of R .

Proof. Consider

$$\text{Rad } R \supseteq (\text{Rad } R)^2 \supseteq \cdots .$$

Assume the chain stops at $B = (\text{Rad } R)^n$. Then $B^2 = B$. If B = 0 we are done. Thus, assume $B \neq 0$. Let A be minimal among the set of left ideals in B with $BA \neq 0$. Then $Ba \neq 0$ for some a in A . Then $Ba \subset A \subset B$ and $B(Ba) = B^2a = Ba \neq 0$. Thus, by the minimality of A , $Ba = A$. Then there is a b in B with $ba = a$. But b is in Rad R and then $b - 1$ is a unit. Thus $a = 0$, which is a contradiction. It is easy to see that Rad R is the largest nilpotent ideal.

An ideal P of R is called _prime_ if any of the following equivalent conditions is satisfied:

(a) If A and B are left (right or two-sided) ideals of R with $AB \subseteq P$ then $A \subseteq P$ or $B \subseteq P$;

(b) If a and b are in R with $aRb \subseteq P$ then a is in P or b is in P ;

(c) If (a) and (b) are principal left (right or two-sided) ideals of R and $(a)(b) \subseteq P$ then $(a) \subseteq P$ or $(b) \subseteq P$.

An element a of R is _strongly nilpotent_ if every sequence

$$a_0, a_1, a_2, \ldots$$

with $a = a_0$ and a_{n+1} in a_nRa_n is ultimately zero. Note if a is strongly nilpotent then a is nilpotent; however, the converse is false. Consider $M_n(GF(q))$ and the standard elementary matrix E_{in} . Then $E_{in}^2 = 0$ but it is easy to select a sequence $A_0 = E_{in}$; $A_1, A_2, \ldots$ with A_{i+1} in $A_iM_n(GF(q))A_i$ which is never ultimately zero.

We have the following characterizations of Rad R which we leave as exercises.

<u>(IV.8)</u> <u>PROPOSITION</u>.

(a) The Rad R is the smallest ideal K such that R/K has only 0 as a nilpotent ideal.

(b) Rad R = $\bigcap P$ where P is a prime ideal of R .

(c) Rad R = $\{r \in R \mid r$ is strongly nilpotent$\}$.

<u>(IV.1)</u> <u>EXERCISE</u>. Prove Proposition (IV.8).

<u>(IV.2)</u> <u>EXERCISE</u>. Let R[X] denote the polynomial ring obtained by adjoining an indeterminate X which commutes with all elements of R . Show(*)

(a) Rad(R[X]) = (Rad R)[X] ,

b) Rad(R[X]) $\cap$ R = Rad R .

(*)Recall R[X] is an exception to the rule that all rings are assumed to be finite. However, we may define the radical of R[X] in the same fashion as in the beginning of this section, i.e., Rad(R[X]) is the intersection of all maximal right ideals of R[X] . Propositions (IV.2), (IV.3), (IV.4) and (IV.5) follow.

(IV.3) EXERCISE. Determine the radical Rad(R) and R/Rad(R) for the following rings:

(a) $\left\{ \begin{bmatrix} a & b \\ c & d \end{bmatrix} \,\middle|\, a,b,c,d \text{ in } GF(q) \right\}$

(b) $\left\{ \begin{bmatrix} a & b \\ 0 & c \end{bmatrix} \,\middle|\, a,b,c \text{ in } GF(q) \right\}$

(c) $\left\{ \begin{bmatrix} a & b \\ 0 & \sigma(a) \end{bmatrix} \,\middle|\, a,b \text{ in } GF(q)\,,\ \sigma \text{ a fixed automorphism} \right\}$

(d) Z/Zp^n (p prime)

(e) $\left\{ \begin{bmatrix} a & b \\ c & d \end{bmatrix} \,\middle|\, a,b,c,d \text{ in } Z/Zp^n \right\}$

(f) $(Z/Zp^n)[X]/(f)$ where f is a monic polynomial which is irreducible modulo (p).

(IV.4) EXERCISE. Let $k = GF(q)$ and $k^* = \langle a \rangle$.

(a) Determine the radical of the ring R generated by

$\begin{bmatrix} 0 & 1 \\ 1 & 0 \end{bmatrix}$ and $\begin{bmatrix} a & 0 \\ 0 & 0 \end{bmatrix}$ in the ring of 2 x 2 matrices over k .

(b) If the characteristic of k is not 2 , determine the radical of the ring R generated by

$\begin{bmatrix} 1 & 0 \\ 0 & 1 \end{bmatrix}$ and $\begin{bmatrix} 1 & 1 \\ 0 & 2 \end{bmatrix}$ in the ring of 2 x 2 matrices over k .

(IV.5) EXERCISE. Show that Rad(R) forms a group under o where $xoy = x + y + xy$. Show that $xoy = 0$ if and only if $(1 - x)(1 - y) = 1$.

(IV.6) EXERCISE. Let R denote the m x m matrices over Z/Zp^n (p prime) and $E_t = \text{diag}(I_t, 0)$ for $0 \leq t \leq m$. Show that $\text{Rad}(E_t R E_t) = E_t \text{Rad}(R) E_t$.

(IV.7) EXERCISE. Let $k = GF(q)$ and consider the skew-polynomial ring $k[X;\sigma]$ of polynomials over k where σ is an automorphism of k and

$$rX = X\sigma(r)$$

for r in k . Let f be a polynomial which generates an ideal in $k[X;\sigma]$. If $R = k[X;\sigma]/(f^n)$, determine Rad(R) and R/Rad(R) . What is $\text{Rad}(k[X;\sigma])$?

(IV.8) EXERCISE. Let R be a ring with $\text{Rad}(R) = N$. If R contains a subring S such that $\text{Rad}(S) = 0$ and S maps under $R \to R/N$ onto R/N then R is called cleft. Otherwise, R is called uncleft. We say that R is maximally uncleft if R/A is uncleft for every ideal $A \subset N$ with $A \neq N$.

(a) Show the following rings are cleft.

(1) $R = R_1 \oplus \cdots \oplus R_n$ where each R_i is cleft.

(2) $(Z/Zp)[X]/(f^n)$ where f is irreducible in $(Z/Zp)[X]$.

(b) Show the following rings are maximally uncleft.

(1) Z/Zp^n .

(2) If $k = GF(q)$, $q = p^t$, then $k[X]/(f)$ where $f = (X^{p^e} - c)^n$, $n > 1$, $e > 0$ and $X^{p^e} - c$ is irreducible over k .

(3) $(Z/Zp^n)[X]/(f)$ where $n > 1$, f is a monic polynomial in $(Z/Zp^n)[X]$ and is irreducible modulo pZ/Zp^n .

V. NAKAYAMA'S LEMMA AND LOCAL RINGS

This section begins with a particularly useful result - Nakayama's Lemma. Various forms and applications of this theorem appear repeatedly in these notes.

A local ring is defined in this section and it is proven that projective modules over local rings are free. This result permits a simple and motivating example concerning the form of idempotents in matrix rings over local rings. This example begins the section on idempotents. The local ring will appear in numerous settings in the structure theory of finite rings. For example, we show in the next section that any finite commutative ring is uniquely expressible as a direct sum of commutative local rings. Later we show that any finite ring is a homomorphic image of a subring of a matrix ring over a local ring.

A ring R is called

(1) local or completely primary if R/Rad(R) is a finite field;

(2) Semi-local or primary if R/Rad(R) is a complete matrix ring over a finite field;

(3) basic is R/Rad(R) is a direct sum of finite fields.

The following is straightforward.

(V.1) THEOREM. The following are equivalent:

(a) R is a local ring.

(b) R has exactly one maximal right (or left) ideal.

(c) The non-units of R are contained in a proper ideal.

(d) The non-units of R form a right (or left) ideal.

(e) The non-units of R form an additive Abelian group.

(f) For every r in R , either r or 1 + r is a unit.

Examples of local rings are

(1) any finite field;

(2) Z/Zp^n (p a prime);

(3) the ring of matrices

$$\left\{ \begin{bmatrix} a & b & c \\ 0 & a & d \\ 0 & 0 & a \end{bmatrix} \middle| \ a,b,c,d \text{ in } Z/Zp \right\} .$$

The first two examples are commutative while the last is non-commutative. To require that R is local is stronger than asking that R have a unique maximal two-sided ideal. To illustrate this let $S = Z/Zp^n$ (p prime) and let $R = M_n(S)$ be the $n \times n$ matrix ring over S $(n \geq 2)$. Then R has a two-sided unique maximal ideal $m = M_n(pZ/Zp^n)$ but

$$R/m = M_n(Z/Zp^n)/M_n(pZ/Zp^n) \simeq M_n(Z/Zp) .$$

Thus R is semi-local but not local.

(V.2) THEOREM. (Nakayama's Lemma) The following conditions on a left ideal J of R are equivalent:

(a) $J \subseteq \mathrm{Rad}(R)$.

(b) Let M be a finitely generated R-module. If $JM = M$ then $M = 0$.

(c) Let M be a finitely generated R-module and N a submodule of M. If $M = N + JM$ then $M = N$.

(d) $1 + J$ is a subgroup of the units R^* of R.

Proof. Clearly (a) and (d) are equivalent. We first show (a) implies (b). Suppose $M \neq 0$. Then select an R-generating set $\{m_1, \ldots, m_n\}$ of M where n is minimal. If $JM = M$ then

$$Jm_1 + \cdots + Jm_n = Rm_1 + \cdots + Rm_n$$

and, in particular, for some s_i in J and r_i in R

$$s_1m_1 + \cdots + s_nm_n = m_1 + r_2m_2 + \cdots + r_nm_n .$$

Thus

$$(1 - s_1)m_1 = (s_2 - r_2)m_2 + \cdots + (s_n - r_n)m_n .$$

But, since s_1 is in J, $1 - s_1$ is invertible and m_1 is in $Rm_2 + \cdots + Rm_n$ which contradicts the minimality of n. To show (b) implies (c), note that $J(M/N) \simeq (JM + N)/N = M/N$ and apply (b). To show (c) implies (d), take s in J and set $r = 1 + s$. Then $R = J + Rr$. By (c), $Rr = R$ and r is a unit.

It is interesting to note that Nakayama's Lemma gives a simple proof that $\text{Rad}(R)$ is nilpotent. Since R is finite, for some n, $\text{Rad}(R)^n = \text{Rad}(R)^{n+1}$. But then if $M = \text{Rad}(R)^n$, $\text{Rad}(R)M = M$ and $M = 0$.

Let R be a commutative ring and A an R-algebra[1] which is finitely generated as an R-module. For an ideal J of R, recall

$$JA = \{ \sum_{\text{finite}} ra \mid r \text{ in } J \text{ and } a \text{ in } A\} .$$

Let $\bigcap (mA)$ denote the intersection of all mA as m runs through the maximal ideals m of R.

<u>(V.3)</u> <u>COROLLARY</u>. Let R be a commutative ring and A an R-algebra which is finitely generated as an R-module. Then,

(a) $\text{Rad}(R)A \subseteq \bigcap (mA) \subseteq \text{Rad}(A)$.

(b) There is an integer n with $\text{Rad}(A)^n \subseteq \bigcap(mA)$.

---

[1] A is an <u>R-algebra</u> if A is a ring and A possesses an R-module structure which is compatible with the ring structure in the obvious fashion.

(c) If A is a projective R-module then $\text{Rad}(R)A = \cap (mA)$.

Proof. Clearly $\text{Rad}(R)A \subseteq \cap (mA)$ since $\text{Rad}(R)A \subseteq mA$ for every maximal ideal m of R . To show the second inclusion suppose that $\cap (mA) \not\subseteq \text{Rad}(A)$. Then there is a maximal left ideal M of A with $mA + M = A$ for every maximal ideal m of R . Hence $M = A$, a contradiction. Statement (b) is trivial since $\text{Rad}(A)$ is nilpotent. To show (c) note that if A is R-free as a module then $\text{Rad}(R)A = \cap (mA)$. Then use the fact that projective modules are direct summands of free modules.

We note that if, as an R/m-algebra, $\text{Rad}(A/mA) = 0$ for every maximal ideal m of R then $\text{Rad}(A) \subseteq mA$ and hence

$$\text{Rad}(A) = \cap (mA) .$$

(V.4) THEOREM. Let R be a local ring with maximal ideal m . Then every finitely generated projective R-module is R-free.

Proof. Let M be R-projective with R-generating set $\{a_1,\ldots,a_n\}$ where n is minimal. Let F be a free R-module with free R-basis $\{x_1,\ldots,x_n\}$ and let $\sigma : F \to M$ by $\sigma(x_i) = a_i$. We have an exact sequence

$$0 \longrightarrow N \xrightarrow{\alpha} F \xrightarrow{\sigma} M \longrightarrow 0 \qquad N = \ker(\sigma)$$

(exact means that α is injective, $\text{Im}(\alpha) = \ker(\sigma)$ and σ is surjective.

Here α is the inclusion morphism). We claim $N \subseteq mF$. Suppose not. Then let $\sum r_i x_i$ be in N and not in mF . Without loss of generality, we may assume r_1 is a unit. But then

$$\sum r_i a_i = \sigma(\sum r_i x_i) = 0 ,$$

$$a_1 = -r_1^{-1}(\sum_{j \neq 1} r_j m_j)$$

and $\{a_1,\ldots,a_n\}$ is not minimal.

Since M is projective, $F = N \oplus M'$ where $M' \simeq M$ (as R-modules). Then $F = N + M' \subseteq mF + M' \subseteq F$. That is, $F = mF + M'$ and by Nakayama's Lemma $F = M'$. Hence M' , and M , is free.

<u>(V.5)</u> <u>THEOREM</u>. Let R be a local ring with maximal ideal m . Let M be a finitely generated R-module.

(a) A subset $\{u_i\}_{i=1}^n$ of M is a generating set for M if and only if their residue classes $\{\bar{u}_i\}_{i=1}^n$ generate the R/m-vector space M/mM .

(b) A subset $\{u_i\}_{i=1}^n$ of M is a minimal generating set for M if and only if $\{\bar{u}_i\}_{i=1}^n$ form an R/m-basis for M/mM .

(c) Any generating set of M contains a minimal generating set. Further, if $\{u_1,\ldots,u_n\}$ and $\{v_1,\ldots,v_m\}$ are both minimal generating sets for M , then $n = m$ and there is an invertible matrix A over R with

$$A\langle u_1,\ldots,u_n\rangle = \langle v_1,\ldots,v_n\rangle .$$

Proof. Simply set $N = \sum Ru_i$ and use Nakayama's Lemma.

Observe that if R is any ring and M a finitely generated R-module, then (also by Nakayama's Lemma) $\{u_i\}_{i=1}^n$ is a minimal generating set for M if and only if the residue classes of $\{u_i\}_{i=1}^n$ form a minimal generating set for M/Rad(R)M over R/Rad(R) .

Also note if M is a free R-module over a local ring R and N is a direct summand of M then by (V.4) N is R-free and by (V.5) any free R-basis of N may be extended to a free R-basis of M .

(V.1) EXERCISE. Prove (V.1) and (V.5).

(V.2) EXERCISE. Show the following rings are local and determine their maximal ideals and ideal lattices.

(a) Z/Zp^n

(b) $GF(q)[X]/(f^n)$ where f is a monic irreducible polynomial in $GF(q)[X]$

(c) $(Z/Zp)[X,Y]/(X^2,Y^2)$

(d) $(Z/Zp)[X,Y]/(X^3,XY,Y^3)$

(e) $(Z/Zp)[X,Y]/(X^3,Y^3)$

(f) $\left\{ \begin{bmatrix} a & b & c \\ 0 & \sigma(a) & 0 \\ 0 & 0 & \beta(a) \end{bmatrix} \middle| \; a,b,c \text{ in } GF(q)\,;\ \sigma, \beta \text{ fixed automorphisms of } GF(q) \right\}$

(g) The set of upper triangular 3×3 matrices $[a_{ij}]$ over $GF(p^t)$ where

$$a_{i+1,j+1} = a_{ij}^s \qquad (i,j = 1,2)$$

and $s = p^r$ for some fixed $r \leq t$.

(V.3) EXERCISE. Let M be a finitely generated free module over a local ring R . Show that a minimal generating set for M is free. Prove the remarks after (V.5).

(V.4) EXERCISE. (Generalized Nakayama Lemma) Let R be a commutative ring and M a finitely generated R-module. Show that an ideal A of R has the property that $AM = M$ if and only if

$$A + \text{Annih}_R(M) = R .$$

(V.5) EXERCISE. Let R be a commutative ring and M be a finitely generated, projective R-module. Let $T_R(M) = \{\sum_i f_i(m_i) \mid m_i \text{ in } M, f_i \text{ in } \text{Hom}_R(M,R)\}$. Show that $T_R(M) \oplus \text{Annih}_R(M) = R$. (Hint: Use Generalized Nakayama Lemma.)

(V.6) EXERCISE. Let R be a commutative ring. Show the following is equivalent to (V.2) Nakayama's Lemma: Let M and N be finitely generated R-modules, $\sigma : M \to N$ an R-morphism, and J an ideal of

R with $J \subseteq \text{Rad}(R)$. Then, if the induced map $\bar{\sigma} : M/JM \to N/JN$ is surjective, so is σ .

(V.7) EXERCISE. Let R be a local ring and M a free finitely generated R-module. Let the R-morphism $\sigma : M \to M$ be idempotent, i.e., $\sigma^2 = \sigma$. Show there exist submodules N and P of M with $M = N \oplus P$, $\sigma|_N$ = identity and $\sigma|_P$ = zero . (Hint: $N(\sigma) = \{x \text{ in } M \mid \sigma(x) = x\}$ and $P(\sigma) = \{x \text{ in } M \mid \sigma(x) = 0\}$.)

(V.8) EXERCISE. Let R be a local ring with characteristic not a power of 2 . Let M be a free finitely generated R-module and $\sigma : M \to M$ an involution; i.e., $\sigma^2 = i_M$. Show there exist submodules N and P of M with $M = N \oplus P$ and $\sigma = i_N \oplus (-i_P)$. (Hint: For x in M , $x = \frac{1}{2}(x + \sigma(x)) + \frac{1}{2}(x - \sigma(x))$.)

(V.9) EXERCISE. Let R be a local ring with maximal ideal m and residue field $k = R/m$. Let β be such that $m^{\beta} = 0$ and $m^{\beta-1} \neq 0$. Show there is a natural sequence of surjective ring morphisms σ_i ;

$$R = R/m^{\beta} \xrightarrow{\sigma_\beta} R/m^{\beta-1} \xrightarrow{\sigma_{\beta-1}} \cdots \xrightarrow{\sigma_2} R/m = k$$

where $\ker(\sigma_i) = m^{i-1}/m^i$, $2 \leq i \leq \beta$. Show that m^{i-1}/m^i , $2 \leq i \leq \beta$, is naturally a k-vector space. If $\phi_i = |m^{i-1}/m^i|$, $2 \leq i \leq \beta$, and $\phi_1 = |k|$, show $|R| = \prod_{i=1}^{\beta} \phi_i$.

(V.10) EXERCISE. Let R be a local ring with maximal ideal m. Show the following are equivalent.

(a) Every left ideal in R is principal.

(b) m is principal.

(c) $\dim_{R/m}(m/m^2) \leq 1$.

(V.11) EXERCISE. (Nagata) Let M be a finitely generated module over a local ring R. If M has a minimal generating set of n elements, show that there exist free R-modules F_1, F_0 with $n = \dim_R(F_0)$ and $m = \dim_R(F_1)$ and $F_1 \xrightarrow{\sigma} F_0 \xrightarrow{\beta} M \to 0$ exact. Deduce that $M \simeq F_0/\mathrm{Im}(\sigma)$. (The exact sequence is called a finite free presentation of M.) If M has minimal generating set $\{v_1, \ldots, v_n\}$ and F_0 has basis $\{x_1, \ldots, x_n\}$ with β given by $\beta : x_i \to v_i$, $1 \leq i \leq n$, then $N = \mathrm{Im}(\sigma) = \ker(\beta)$ is called the relation module of M. Show that (use (V.5)) the relation module N of M is unique up to R-module isomorphism.

(V.12) EXERCISE. (Continuation of the previous exercise.) (Nagata) We define the notion of the syzygies of the R-module M. The 0-th syzygie of M is M itself; when the i-th syzygie N_i of M is defined then the (i + 1)-th syzygie is the relation module for N_i. Denote the i-th syzygie of M by $\mathrm{syz}^i M$. Then M is said to have homological dimension n if $n + 1$ is the smallest i such that $\mathrm{syz}^i M = 0$, written as $\mathrm{hd}(M) = n$. Show

(a) $\mathrm{hd}(M) = -1$ if and only if $M = 0$.

(b) $\mathrm{hd}(M) = 0$ if and only if M is a free R-module different from zero.

(c) If $M \neq 0$ and F is a free finitely generated R-module, then $hd(M) = hd(F \oplus M)$.

(d) $hd(syz^i M) = \max(-1, hd(M) - i)$.

(syzygy (pronounced sĭz'ĭ·jĭ) is of Greek origin and means to "couple" or to "yoke together".)

(V.13) EXERCISE. (Kaplansky) Let R be a commutative local ring, M a finitely generated R-module and

$$F_1 \xrightarrow{\sigma} F_0 \xrightarrow{\beta} M \longrightarrow 0$$

a finite free presentation of M . A matrix A over R is said to admit diagonal reduction if there exist invertible matrices P and Q such that

$$PAQ = \text{diag}[d_1, d_2, \ldots]$$

where d_i divides d_{i+1} .

(a) Show that if the matrix of σ admits diagonal reduction to $\text{diag}[d_1, d_2, \ldots, d_n]$ then $M \simeq R/Rd_1 \oplus \cdots \oplus R/Rd_n$.

(b) Show that if all 1 x 2 , 2 x 1 , and 2 x 2 matrices over R admit diagonal reduction, then all matrices admit diagonal reduction.

(c) Show that if R is a principal ideal ring then all matrices over R admit diagonal reduction.

(V.14) EXERCISE. (Lenzing) A ring R is called a *left Steinitz ring* if any R-free subset of a free left R-module F can be extended to a free R-basis for F . Show

(a) R is a left Steinitz ring if and only if for any free R-module ${}_RF$, and any free submodule ${}_RH$ then F/H is free and H is a direct summand of F .

(b) Every projective module over a left Steinitz ring is free.

(c) A local ring is a left Steinitz ring.

(V.15) EXERCISE. Let R be a local commutative ring. Let $a_1,\ldots,a_n$ be in R with $R = Ra_1 + \cdots + Ra_n$. Show there exists an n x n matrix A with entries from R such that the first row is $[a_1,\ldots,a_n]$ and det(A) = 1 . [Comment and Hint: Show that this is equivalent to the fact that all finitely generated projective R-modules are free.]

(V.16) EXERCISE. Consider $M = Z/Zp^n \oplus Z/Zp^n$ (n > 1) where p is a prime. The Z/Zp^n-module has a basis $e_1 = [1,0]$ and $e_2 = [0,1]$. Let N be the submodule generated by e_1 and pe_2 .

(a) Show that N has a minimal generating set of two elements and that $N \simeq Z/Zp^n \oplus Z/Zp^{n-1}$.

(b) Show that N/(p)M is a vector space of dimension one in M/(p)M . Does this contradict (V.5)?

(V.17) EXERCISE. Let R be a local ring with maximal ideal m and F a free R-module. Let F have basis $\{b_1,\ldots,b_n\}$ and $\alpha = r_1b_1 + \cdots + r_nb_n$. Show the following are equivalent.

(1) $\Pi\alpha \neq 0$ under $\Pi : F \to F/mF$.

(2) $Rr_1 + \cdots + Rr_n = R$.

(3) r_i is a unit for some i .

(4) $R\alpha$ is an R-direct summand of F .

(5) There exists an R-morphism $\sigma : F \to R$ with $\sigma(\alpha) = 1$.

VI. STRUCTURE THEOREM FOR FINITE COMMUTATIVE RINGS

This section is devoted to showing that any finite commutative ring may be expressed uniquely as a direct sum of finite local commutative rings.

Let R be a ring and let $A_1, \ldots, A_n$ be ideals of R. Two ideals A_i and A_j are called <u>coprime</u> if $A_i + A_j = R$. Define a ring morphism

$$\phi : R \to R/A_1 \oplus \cdots \oplus R/A_n$$

by

$$\phi : r \to \langle r + A_1, \ldots, r + A_n \rangle .$$

It is easy to verify that,

(1) If, whenever $i \neq j$, A_i and A_j are coprime then $A_1 \cap \cdots \cap A_n = A_1 \cdots A_n$.

(2) If A_i and A_j are coprime then A_i^m and A_j^m are coprime for $m = 1,2,3,\ldots$.

(3) The morphism ϕ is injective if and only if $A_1 \cap \cdots \cap A_n = 0$.

(4) The morphism ϕ is surjective if and only if, whenever $i \neq j$, A_i and A_j are coprime.

An element e of a ring R is called an <u>idempotent</u> if $e^2 = e$. Two idempotents e and f are called <u>orthogonal</u> if $ef = 0$ and e is <u>central</u> or <u>commutative</u> if $er = re$ for all r in R. The next result is easy.

<u>(VI.1)</u> <u>PROPOSITION</u>. Let R be a ring. The following statements are equivalent:

(a) R is isomorphic to a direct sum of rings R_i, $1 \leq i \leq n$.

(b) There exist commutative orthogonal idempotents e_i in R with $1 = \sum_{i=1}^{n} e_i$ and $R_i \simeq e_i R$.

(c) R is a direct sum of ideals $A_i \simeq R_i$, $1 \leq i \leq n$.

We can now prove:

<u>(VI.2)</u> <u>THEOREM</u>. (Structure Theorem for Finite Commutative Rings) Let R be a commutative ring. Then R decomposes (up to order of summands) uniquely as a direct sum of local rings.

<u>Proof</u>. Let $P_1, \ldots, P_n$ be the prime ideals of R. Since R/P_i is a finite domain (hence a finite field), $P_1, \ldots, P_n$ are maximal ideals of R. Then $\mathrm{Rad}(R) = P_1 \cap \cdots \cap P_n$. Since, whenever $i \neq j$, P_i and P_j are coprime we have $P_1 \cap \cdots \cap P_n = P_1 \cdots P_n$ and there exists a smallest integer m satisfying

$$(\mathrm{Rad}(R))^m = P_1^m \cdots P_n^m .$$

Define

$$\phi : R \to (R/P_1^m) \oplus \cdots \oplus (R/P_n^m)$$

by

$$\phi : r \to \left\langle r + P_1^m, \ldots, r + P_n^m \right\rangle .$$

Since $\cap P_i^m = 0$, ϕ is injective and since P_i^m and P_j^m $(i \neq j)$ are coprime, ϕ is surjective.

The ideals of R/P_i^m are in a natural correspondence with the ideals of R containing P_i^m . But P_i is the only maximal ideal with $R \supset P_i \supset P_i^m$. Thus R/P_i^m is local with unique maximal ideal P_i/P_i^m .

Suppose we have two decompositions of R . Corresponding to these decompositions are representations of $1 = \sum_i e_i = \sum_j f_j$ as orthogonal idempotents. Since the summands Re_i and Rf_j are local, no e_i is the sum of two proper orthogonal idempotents. Thus, since $e_i = \sum_j e_i f_j$, e_i must equal $e_i f_{j(i)}$. Likewise $f_j = f_j e_{i(j)}$. We obtain

$$e_i = e_i f_{j(i)} = e_i f_{j(i)} e_{i(j(i))}$$

which implies

$$i = i(j(i))$$

since the e_i are orthogonal. Thus, there is a bijection between $\{e_i\}$ and $\{f_j\}$. Hence $e_i = e_i f_{j(i)}$, $f_{j(i)} = f_{j(i)} e_i$ and thus $e_i = f_{j(i)}$.

(VI.1) EXERCISE. Prove the remarks before Proposition (VI.1).

(VI.2) EXERCISE. Prove Proposition (VI.1).

(VI.3) EXERCISE. (Köthe) Let R be a commutative ring and suppose every R-module is a direct sum of cyclic R-modules. Show that every ideal of R is principal.

(VI.4) EXERCISE. Determine the decompositions into local rings of the following commutative rings.

(a) $GF(q)[X]/(f)$ where $f(X) = X^{p-1} + \cdots + X + 1$ and p and q are distinct primes.

(b) $(Z/Zq^n)[X]/(f)$ where f is the same as (a).

(c) $GF(q)[X]/(f)$ where $f = p_1^{t_1} \cdots p_m^{t_m}$ and p_i is irreducible for $1 \leq i \leq m$.

(d) Z/Zm.

(VI.5) EXERCISE. Let $R_1, \ldots, R_n$ be commutative rings and $R = R_1 \oplus \cdots \oplus R_n$.

(a) If R^* denotes the units of R , show $R^* = R_1^* \times \cdots \times R_n^*$.

(b) Show $R[X] = \oplus \sum_{i=1}^{n} R_i[X]$.

(c) If $M_m(R)$ denotes the $m \times m$ matrix ring over R, show $M_m(R) = \oplus \sum_{i=1}^{n} M_m(R_i)$.

(d) If A is an ideal of R, show $A = \oplus \sum_{i=1}^{n} A_i$ where A_i is an ideal in R_i . Further, show A is a maximal ideal if and only if for some i , A_i is a maximal ideal of R_i and $A_j = R_j$ for $j \neq i$.

(e) Show that R has non-zero nilpotents if and only if R_i has non-zero nilpotents for some i .

(f) Show that if $n \geq 2$ then R has zero divisors.

(g) If the characteristic of R_i is m_i , show that R has characteristic $m = \text{l.c.m.}(m_1, \ldots, m_n)$.

(h) If the R_i are non-commutative, which of the above are true?

VII. IDEMPOTENTS

This section provides the basic theory for the next several sections. We will approach the structure of a finite ring R as follows: There exists a natural anti-isomorphism between R and $\mathrm{End}({}_R R)$. Thus to determine R we examine $\mathrm{End}({}_R R)$. But ${}_R R = L_1 \oplus \cdots \oplus L_n$ where L_i are left ideals and consequently $\mathrm{End}({}_R R) \simeq \oplus \sum_{i,j} \mathrm{Hom}_R(L_i, L_j)$. Since each left ideal L_i is of the form Re_i where e_i is an idempotent, much of the problem reduces to an examination of the left ideal Re and the R-morphisms $\mathrm{Hom}_R(Re, Rf)$ for idempotents e and f. We are particularly interested in the case where Re is indecomposable which is equivalent to the ring eRe being local. Thus this section examines the left ideal Re, the ring eRe and its radical, the R-morphisms $\mathrm{Hom}_R(Re, Rf)$, the uniqueness of decompositions of R as left ideals and finally the "lifting" of idempotents from $R/\mathrm{Rad}(R)$ to R.

One might say that presently the philosophy of approach is "matrix theoretic" - thus idempotents play a central role. To illustrate this we begin with an example of idempotents in a matrix ring over a local ring. Indeed, this example is significant since we later show that a finite ring is a homomorphic image of a subring of a matrix ring over a local ring and, of course, idempotents are preserved under ring morphism.

An element e in a ring R is called an idempotent if $e^2 = e$. Two idempotents are called orthogonal if $ef = fe = 0$. An idempotent e is called central or commutative if $re = er$ for all r in R and non-trivial if $e \neq 0,1$.

Important Example. Let R be a local ring and let $R^{(n)}$ denote a free R-module with R-basis of n elements. Suppose that $e : R^{(n)} \to R^{(n)}$ is an R-morphism satisfying $e^2 = e$; i.e., e is an idempotent in $\mathrm{End}_R(R^{(n)})$.

Let
$$N(e) = \{x \text{ in } R^{(n)} \mid e(x) = x\}$$
and
$$P(e) = \{x \text{ in } R^{(n)} \mid e(x) = 0\}.$$

Clearly $N(e) \cap P(e) = 0$ and for x in $R^{(n)}$, $x = e(x) + (x - e(x))$ where $e(x)$ is in $N(e)$ and $x - e(x)$ is in $P(e)$. Thus $R^{(n)} = N(e) \oplus P(e)$. Consequently, both $N(e)$ and $P(e)$ are projective and, by (V.4), free. Selecting free R-bases for $N(e)$ and $P(e)$ we obtain a free R-basis for $R^{(n)}$ and relative to this basis e has the matrix representation

$$\mathrm{Mat}(e) = \begin{bmatrix} I_t & 0 \\ 0 & 0 \end{bmatrix}$$

where I_t is a $t \times t$ identity matrix and $t = \dim_R(N(e))$.

If $S = \mathrm{End}_R(R^{(n)})$ then

$$Se \simeq \left\{ \begin{bmatrix} a_{11} & \cdots & a_{1t} & 0 & \cdots & 0 \\ \\ a_{n1} & \cdots & a_{nt} & 0 & \cdots & 0 \end{bmatrix} \;\middle|\; a_{ij} \text{ in } R \right\}$$

(as left S-module) and

$$eSe \simeq \left\{ \begin{bmatrix} A & 0 \\ \\ 0 & 0 \end{bmatrix} \;\middle|\; A \text{ in } M_t(R) \right\}$$

(as a ring). In particular, eSe is local if and only if e has

$$\mathrm{Mat}(e) = \begin{bmatrix} 1 & 0 & \cdots & 0 \\ 0 & 0 & \cdots & 0 \\ \cdot & \cdot & & \cdot \\ \cdot & \cdot & & \cdot \\ 0 & 0 & \cdots & 0 \end{bmatrix}$$

and here $eSe \simeq R$ (as a ring). This completes the example.

If e is an idempotent in R then $1 - e$ is an idempotent orthogonal to e and

$$1 = e + (1 - e)$$

gives a direct sum decomposition of ${}_RR$ as left ideals (or as left R-modules)

$${}_RR = Re \oplus R(1 - e) .$$

Conversely, if for left ideals L_1 and L_2 we have ${}_RR = L_1 \oplus L_2$, then $1 = e_1 + e_2$ where e_1 is in L_1 and e_2 is in L_2. Then $e_i^2 = e_i$ $(i = 1,2)$ and $e_1e_2 = e_2e_1 = 0$. Further $L_1 = Re_1$ and $L_2 = Re_2$. A left ideal L which is a direct summand of ${}_RR$ is called a <u>component</u> of R.

Hence a left ideal L is a component if and only if $L = Re$ for some idempotent e of R.

Much of what is to be developed depends on the R-endomorphisms of R or of a component of ${}_RR$. We pause to examine endomorphisms of finite direct su

Suppose a left R-module M is given as a direct sum

$$M = M_1 \oplus \cdots \oplus M_m$$

of left R-modules M_i , $1 \le i \le m$.

Let $\lambda_j : M_j \to M$ be the natural inclusion and $\Pi_j : M \to M_j$ the natural projection for $1 \le j \le m$. If x is in M then

$$x = x_1 + \cdots + x_m \qquad \text{(uniquely)}$$

where x_i is in M_i . Further

$$\Pi_j x = x_j ,$$

$$\Pi_i \lambda_j \Pi_j x = 0 \qquad \text{if } i \neq j ,$$

and

$$\sum_j \lambda_j \Pi_j x = x ,$$

i.e.,

$$\sum_j \lambda_j \Pi_j = i_M .$$

Suppose $\sigma : M \to M$ is an R-endomorphism. Set

$$\sigma_{ij} = \Pi_i \sigma \lambda_j$$

and note $\sigma_{ij} : M_j \to M_i$. In this fashion we may associate σ to a matrix $\mathrm{Mat}(\sigma)$ of the σ_{ij} ,

$$\mathrm{Mat} : \sigma \to \mathrm{Mat}(\sigma) = \begin{bmatrix} \sigma_{11} & \sigma_{12} & \cdots & \sigma_{1m} \\ \sigma_{21} & \sigma_{22} & \cdots & \sigma_{2m} \\ \vdots & \vdots & & \vdots \\ \sigma_{m1} & \sigma_{m2} & \cdots & \sigma_{mm} \end{bmatrix}$$

$$= [\sigma_{ij}]$$

where $\sigma_{ij} = \Pi_i \sigma \lambda_j : M_j \to M_i$.

<u>(VII.1)</u> <u>THEOREM</u>. Let $M = M_1 \oplus \cdots \oplus M_m$ (R-modules). Then Mat determines a ring isomorphism

$$\mathrm{Mat} : \mathrm{End}_R(M) \to \{[\sigma_{ij}] \mid \sigma_{ij} : M_j \to M_i\}$$

defined by

$$\mathrm{Mat} : \sigma \to \mathrm{Mat}(\sigma) = [\sigma_{ij}]$$

where

$$\sigma_{ij} = \Pi_i \sigma \lambda_j \ .$$

<u>Proof</u>. Suppose $\psi : M \to M$ with $\mathrm{Mat}(\psi) = [\psi_{ij}]$. Then, since

$$(\sigma + \psi)_{ij} = \Pi_i(\sigma + \psi)\lambda_j = \Pi_i \sigma \lambda_j + \Pi_i \psi \lambda_j = \sigma_{ij} + \psi_{ij} ,$$

we have

$$\mathrm{Mat}(\sigma + \psi) = \mathrm{Mat}(\sigma) + \mathrm{Mat}(\psi) \ .$$

While
$$\begin{aligned}(\sigma\psi)_{ij} &= \Pi_i(\sigma\psi)\lambda_j \\ &= \Pi_i\sigma(\textstyle\sum_k\lambda_k\Pi_k)\psi\lambda_j \\ &= \textstyle\sum_k (\Pi_i\sigma\lambda_k)(\Pi_k\psi\lambda_j) \\ &= \textstyle\sum_k \sigma_{ik}\psi_{kj} ,\end{aligned}$$

implies $\mathrm{Mat}(\sigma\psi) = \mathrm{Mat}(\sigma)\ \mathrm{Mat}(\psi)$. This gives the fact that Mat is a ring morphism. It is easy to show that it is bijective.

We utilize (VII.1) in the following fashion. Since R is finite, it is an easy exercise to show that

$$_RR = L_1 \oplus \cdots \oplus L_n$$

where the L_1 are indecomposable left ideals. By the Krull-Schmidt Theorem this decomposition is unique to within isomorphism. But

$$\mathrm{End}_R(_RR) \simeq R^0 \qquad \text{(as rings)}$$

where R^0 denotes the opposite ring of R . While on the other hand, by (VII.1) $\mathrm{End}_R(_RR)$ is isomorphic to the ring of matrices $[\sigma_{ij}]$ where $\sigma_{ij} : L_j \to L_i$ is an R-morphism. But L_i is a direct summand of $_RR$; i.e., a component of R , thus $L_i = Re_i$ is generated by an idempotent e_i . Hence, for idempotents e and f , we find we need to examine $\mathrm{Hom}_R(Re,Rf)$.

Suppose M is a left R-module and e is an idempotent of R. If $\rho : Re \to M$ is an R-morphism then $\rho(e) = m$ for some m in M. Indeed, since $e^2 = e$ we have $m = em$ since $\rho(e) = \rho(e^2) = e\rho(e)$. Then if r is in Re, $r = re$ and

$$\rho(r) = \rho(re) = r\rho(e) = rem .$$

We denote ρ by ρ_{em} and the map $\rho_{em} \to em$ gives a mapping $\mathrm{Hom}_R(Re,M) \to eM$.

Conversely, given em in eM the map $e \to em$ determines an R-morphism $\rho_{em} : Re \to M$ by $\rho_{em}(r) = rem$. Observe, this is well-defined since if $r_1 = r_2$ (r_i in Re) then $r_1e = r_2e$ and $r_1em = r_2em$. This gives a natural map $\beta : eM \to \mathrm{Hom}_R(Re,M)$ where $\beta : em \to \rho_{em}$.

Clearly the maps are bijective and additive. Thus $\mathrm{Hom}_R(Re,M) \simeq eM$ (as additive groups).

<u>(VII.2)</u> <u>THEOREM</u>. Let e and f be idempotents in R and let M be a left R-module. Then

(a) $\mathrm{Hom}_R(Re,M) \simeq eM$ (as groups).

(b) $\mathrm{Hom}_R(Re,Rf) \simeq (eRf)$ (as groups).

Further, Re is R-isomorphic to Rf if and only if there is an element c in eRf and d in fRe with

$$cd = e \quad \text{and} \quad dc = f .$$

Thus, by symmetry, $Re \simeq Rf$ (as R-modules) if and only if $eR \simeq fR$ (as R-modules).

(c) $End_R(Re) \simeq (eRe)^\circ$ (as rings).

(d) If S is a <u>right</u> ideal of R, then an R-isomorphism $Re \to Rf$ maps $S \cap Re \to S \cap Rf$.

<u>Proof</u>. We have (a). Suppose $\rho : Re \to Rf$ is an R-isomorphism. Then for some c in eRf, $\rho = \rho_c$. Let d be in Re with $\rho_c : d \to f$. Then $f = \rho_c(d) = dc$. Now $\rho : e \to c$ thus $\rho^{-1} : c \to e$. But $\rho^{-1} : f \to d$. Thus d is in fRe and $e = \rho^{-1}(c) = \rho^{-1}(cf) = c\rho^{-1}(f) = cd$.

Now suppose $cd = e$ and $dc = f$. We may assume that c is in eRf and d is in fRe, since

$$\begin{aligned}(ecf)(fde) &= ecfde \\ &= ecdcde \\ &= e^4 = e\end{aligned}$$

and similarly $(fde)(ecf) = f$. Then, since $cd = e$ and $dc = f$ it is easy to see that

$$Re \to Rf \quad \text{by} \quad x \to xc$$

and

$$Rf \to Re \quad \text{by} \quad y \to yd$$

are R-isomorphisms.

The remaining parts of the theorem are straightforward.

<u>We restrict our attention to a special case</u>:

Suppose

$$_R R = L_1 \oplus \cdots \oplus L_n$$

where the L_i are left ideals and $L_i \simeq L_j$ (as R-modules) for all i and j.

Then $L_i = Re_i$, $1 \leq i \leq n$, $\{e_i\}$ are orthogonal idempotents and if e denotes one of the e_i there exist c_j and d_j with

$$c_j d_j = e \qquad d_j c_j = e_j$$

for each j, $1 \leq j \leq n$.

By the orthogonality of $\{e_i\}$,

$$eR = eRe_1 \oplus \cdots \oplus eRe_n$$

(as a left eRe-module). Moreover

$$ere_j \to ere_j d_j = erd_j c_j d_j = erd_j e$$

is easily seen to be an eRe-isomorphism $eRe_j \to eRe$. That is, eR is a free eRe-module on n-generators.

(VII.3) THEOREM. Let

$${}_RR = L_1 \oplus \cdots \oplus L_n$$

where the L_i are left ideals with $L_i \simeq L_j$ (as left R-modules) for all i and j. Then there exists idempotents e_i with $L_i = Re_i$, $1 \leq i \leq n$, and if e denotes one of the e_i,

(a) eR is a free eRe-module with free eRe-basis of n-elements.

(b) $R^0 \simeq \mathrm{End}_R({}_RR) \simeq \mathrm{End}_{eRe}(eR)$.

(c) Consequently, R^0 is isomorphic (as a ring) to the $n \times n$ matrix ring over $(eRe)^\circ$

The proof is clear. (Observe $\mathrm{End}_R(\oplus \sum Re) \simeq M_n((eRe)^\circ) \simeq \mathrm{End}_{eRe}(eR)$.)

The purpose of the remainder of this section is an examination of the left ideal Re and the ring eRe where e is an idempotent.

(VII.4) THEOREM. Let e be a non-zero idempotent of R. Then

$$\mathrm{Rad}(eRe) = eRe \cap (\mathrm{Rad}(R)) = e\mathrm{Rad}(R)e .$$

Proof. We show $\mathrm{Rad}(eRe) \subseteq eRe \cap (\mathrm{Rad}(R)) \subseteq e\mathrm{Rad}(R)e \subseteq \mathrm{Rad}(eRe)$.

Suppose that $r = ere$ is in $\mathrm{Rad}(eRe)$ and x is an element of R. Then $e = rxe$ has inverse $e - y$ in eRe where $y = eye$. Then $(e - rxe)(e - y) = e$ and $(1 - rx)(1 - y) = 1 - rx(1 - e)$. Now multiplication on the right by $1 - rx(1 - e)$ gives 1 ; i.e., $1 - rx$ is invertible and hence $Rad(eRe) \subseteq (eRe) \cap (Rad(R))$.

Suppose r is in $eRe \cap Rad(R)$. Then $r = ere$ and r is in $Rad(R)$. That is,

$$(eRe) \cap Rad(R) \subseteq e(Rad(R))e \ .$$

Finally, assume that r is in $e(Rad(R))e$ and thus in $Rad(R)$. Then for $x = exe$ in eRe the element $1 - rx$ has inverse $1 - y$ in R . Then, in eRe ,

$$(e - rx)(e - eye) = e(1 - rx)(1 - y)e = e$$

and $e(Rad(R))e \subseteq Rad(eRe)$.

<u>(VII.5)</u> <u>THEOREM</u>. Let L be a non-nilpotent ideal in R . Then L contains a non-zero idempotent.

<u>Proof</u>. We use induction on the order of L , $|L|$. If $|L| = 2$ it is trivial. Thus assume the theorem is true for left ideals of order $< n$.

Let $|L| = n$. Suppose $La = L$ for some a in L . Since L is finite $xa = 0$ only if $x = 0$ for x in L . Also $ea = a$ for some $e \neq 0$ in L . Then $(e^2 - e)a = 0$ and $e^2 - e = 0$.

Suppose $|La| < |L|$. By induction either La is nilpotent or it contains an idempotent. If La is nilpotent for each a in L then each a in L is itself nilpotent.

We claim that L is nilpotent if each a in L is nilpotent. Suppose L is not nilpotent. Select a left ideal J in L which is smallest relative to being not nilpotent. Then $J^2 \subseteq J$ and J^2 is not nilpotent. Hence $J^2 = J$. Now let H be the smallest left ideal in R with $JH \neq 0$. Select h in H with $Jh \neq 0$. Then $J(Jh) = Jh \neq 0$ and $Jh \subset H$. Thus $Jh = H$. Finally, select t in J with $th = h$. Then

$$t^m h = t^{m-1}h = \cdots = th = h \neq 0$$

for each positive integer m . But t is in $J \subseteq L$ and is nilpotent - a contradiction since $h \neq 0$. But then L is nilpotent - a contradiction to the hypothesis.

Similar arguments will now prove the corollary.

<u>(VII.6)</u> <u>COROLLARY</u>. Let L be a minimal left ideal of R . Then $L^2 = 0$ or $L = Re$ for an idempotent e of R .

<u>(VII.7)</u> <u>THEOREM</u>. The following are equivalent:

(a) R is local.

(b) R contains no non-trivial idempotents.

Proof. Clearly (a) implies (b). Assume (b) and suppose that a is in R and not in $Rad(R)$. Then Ra is not nilpotent. Thus Ra contains an idempotent e. By (b), $e = 1$. Thus, there is a b in R with $ba = 1$ and thus $R/Rad(R)$ is a finite field.

For idempotents e and f we say $e > f$ if $ef = fe = f$; equivalently, f is in eRe. If $B(R)$ denotes the idempotents of R then $B(R)$ is partially ordered under $\geq$ with smallest element 0 and maximal element 1. A smallest non-zero idempotent under $\geq$ is called a minimal idempotent.

(VII.8) THEOREM. Let e be non-zero idempotent of R.

(a) The following are equivalent.

(1) e is minimal.

(2) eRe is a local ring.

(3) Re is indecomposable.

(b) If $Rad(R) = 0$, the following are equivalent.

(1) Re is a minimal left ideal.

(2) eRe is a finite field.

Proof. For part (a) it is clear that (1) is equivalent to (2). To show (3) implies (2) suppose Re is indecomposable and eRe is not local. Let f be a non-zero idempotent in eRe with $f \neq e$. Then f and $e - f$ are orthogonal idempotents in eRe whose sum is e. Hence Re is decomposable.

To show (2) implies (3) suppose eRe is local. Suppose also that $e = e_1 + e_2$, $e_1 e_2 = 0$ and e_i are non-zero. Then $e = e^2 = ee_1 + ee_2$. Subtract $e = e_1 + e_2$ and we have $e_1 + e_2 - ee_1 - ee_2 = 0$. This implies e_1 and e_2 are in eR. Similarly e_1 and e_2 are in Re and hence in eRe which contradicts the assumption that eRe is local.

For part (b) if Re is minimal then by Schur's Lemma $End_R(Re) = eRe$ is a division ring (hence a finite field). Conversely, suppose $End_R(Re) = eRe$ is a finite field. Let r be in R with $re \neq 0$ and consider the ideal Rre. Since $Rad(R) = 0$ we have $(re)R(re) \neq 0$ and thus there is an s with $esre \neq 0$. But $esre$ is in eRe and thus has an inverse ete in eRe. Then $((ete)es)re = e$ so that $Re \subset Rre$. Clearly $Rre \subseteq Re$. Hence $Re = Rre$ and Re is minimal.

<u>(VII.9)</u> <u>THEOREM</u>. Let e be a minimal non-zero idempotent of R. Let $\Pi : R \to R/Rad(R)$ be the natural morphism. Then ΠRe is a minimal left ideal.

<u>Proof</u>. Consider $\Pi R(1 - e)$ in $R/Rad(R)$. This is a proper ideal and thus contained in a maximal left ideal of the form ΠM in $R/Rad(R)$. Suppose $\Pi M \cap \Pi Re \neq 0$. Then $(\Pi M \cap \Pi Re)^2 \neq 0$ and there exists an r in R with $\Pi(re)$ in ΠM and $\Pi(ere) \neq 0$. But $\Pi(eRe)$ is a finite field since $\Pi(eRe) = eRe/Rad(eRe)$ and eRe is local. Thus there is a Πx with $\Pi(xere) = \Pi e$. Hence Πe is in ΠM. Thus e is in M

and $1 = e + (1 - e)$ is in M - a contradiction. Hence $\Pi M \cap \Pi Re = 0$ and ΠRe is minimal.

Let ${}_R R = Re_1 \oplus \cdots \oplus Re_n$ where the Re_i are indecomposable left ideals. Set

$$N_i = Re_i \cap \mathrm{Rad}(R) .$$

Then N_i is a greatest submodule of Re_i since $Re_i/\mathrm{Rad}(R)$ is a minimal ideal.

<u>(VII.10)</u> <u>THEOREM</u>. Let ${}_R R = Re_1 \oplus \cdots \oplus Re_n$ where the Re_i are indecomposable. Set $N_i = Re_i \cap \mathrm{Rad}(R)$. Then $Re_i \simeq Re_j$ if and only if $Re_i/N_i \simeq Re_j/N_j$.

<u>Proof</u>. If $Re_i \simeq Re_j$ the statement is clear. Conversely, let

$$\omega : Re_i/N_i \to Re_j/N_j$$

be an isomorphism and let $\Pi_i : Re_i \to Re_i/N_i$. We have

$$\begin{array}{ccccc} & & Re_i & & \\ & & \downarrow{\scriptstyle \omega\Pi_i} & & \\ Re_j & \xrightarrow{\Pi_j} & Re_j/N_j & \longrightarrow & 0 \end{array} \qquad \text{(exact row).}$$

Since Re_i is projective there is a $\phi : Re_i \to Re_j$ with $\Pi_j\phi = \omega\Pi_i$. Note that $\omega\Pi_i \neq 0$ so $\Pi_j\phi \neq 0$. Thus $\phi(Re_i) \not\subseteq \ker\Pi_i = N_i$. Thus

$\phi(Re_i) = Re_j$. By symmetry, $Re_i \simeq Re_j$.

To conclude our discussion of idempotents we would like to illustrate how idempotents in R may be obtained from idempotents in R/Rad(R) . We say that u is an <u>idempotent modulo Rad(R)</u> if $u^2 - u$ is in Rad(R) and that u may be <u>lifted</u> if there is an idempotent e in R with e - u in Rad(R) .

Suppose that u is an idempotent modulo Rad(R) . Set $x_1 = u^2 - u$. If $x_1 = 0$ then u is an idempotent. If $x_1 \neq 0$ set

$$e_1 = u + x_1 - 2ux_1 .$$

Then $e_1^2 - e_1 = 4x_1^3 - 3x_1^2 = x_1^2(4x_1 - 3)$. But x_1 is in Rad(R) and is thus nilpotent. Further $e_1 \equiv u \bmod(\mathrm{Rad}(R))$. Then either e_1 is idempotent or we may repeat the process. Eventually we arrive at an idempotent e with $e \equiv u \bmod(\mathrm{Rad}(R))$.

<u>(VII.11)</u> <u>THEOREM</u>. Idempotents modulo the radical may be lifted.

Let S denote R/Rad(R) . Then ${}_S S$ decomposes

$${}_S S = \overline{L}_1 \oplus \cdots \oplus \overline{L}_n$$

as a direct sum of indecomposable ideals. Since Rad(S) = 0 , $\overline{L}_i \cap \mathrm{Rad}(S) = 0$ and the $\overline{L}_i$ are minimal. Thus we have a decomposition

$${}_S S = S\bar{e}_1 \oplus \cdots \oplus S\bar{e}_n$$

where the $\bar{e}_i$ are minimal mutually orthogonal idempotents with $\bar{e}_1 + \cdots + \bar{e}_n = \bar{1}$.

<u>(VII.12)</u> <u>THEOREM</u>. Let $S = R/\text{Rad}(R)$ decompose

$$_S S = S\bar{e}_1 \oplus \cdots \oplus S\bar{e}_n$$

where the $\bar{e}_i$ are minimal mutually orthogonal idempotents. Then

$$_R R = Re_1 \oplus \cdots \oplus Re_n$$

where the e_i are local mutually orthogonal idempotents and $e_i \equiv \bar{e}_i \bmod(\text{Rad}(R))$. Conversely, a decomposition of $_R R$ of the above form implies a decomposition of $_S S$.

<u>Remark</u>: Observe that the following proof also shows that if $\bar{1} = \bar{e}_1 + \cdots + \bar{e}_n$ where the $\bar{e}_i$ are mutually orthogonal idempotents in $R/\text{Rad}(R)$ then $1 = e_1 + \cdots + e_n$ in R where the e_i are mutually orthogonal idempotents with $e_i \equiv \bar{e}_i$ (mod Rad(R)) .

<u>Proof</u>. The proof is by induction. We may assume local mutually orthogonal idempotents $e_1,\ldots,e_{s-1}$ have been obtained from $\bar{e}_1,\ldots,\bar{e}_{s-1}$, respectively. Set $f = e_1 + \cdots + e_{s-1}$ and let

$$\lambda = (1 - f)u(1 - f)$$

where u is in R and $u \equiv \bar{e}_s \bmod(\text{Rad}(R))$. Then $\lambda \equiv \bar{e}_s \bmod(\text{Rad}(R))$

and $e_i\lambda = \lambda e_i = 0$ for $i < s$. If λ is an idempotent we are done. If not, set $x_1 = \lambda^2 - \lambda$, x_1 is in $\mathrm{Rad}(R)$, and repeat procedure outlined prior to (VII.11). We arrive at an idempotent e_s with $e_s \equiv \bar{e}_s \bmod(\mathrm{Rad}(R))$ and $e_s e_i = e_i e_s = 0$ for $i < s$. It is easy to see that e_s is indecomposable and hence local. It is also straightforward to show that if $1 = \bar{e}_1 + \cdots + \bar{e}_n$ then $1 = e_1 + \cdots + e_n$.

After (VII.1) it was noted that a decomposition ${}_R R = Re_1 \oplus \cdots \oplus Re_n$ where the e_i are local was unique up to isomorphism by the Krull-Schmidt Theorem. Without appealing to the Krull-Schmidt Theorem we could show that if R is semi-simple; i.e., $\mathrm{Rad}(R) = 0$, then ${}_R R$ decomposes uniquely as a direct sum of minimal left ideals. We show that this is sufficient to prove uniqueness of the decomposition for an arbitrary ring R. Thus, we provide a second approach to the Krull-Schmidt Theorem.

<u>(VII.13)</u> <u>THEOREM</u>. Let $1 = \sum_{i=1}^m e_i$ and $1 = \sum_{j=1}^n f_j$ where e_i and f_j are orthogonal local idempotents. Then $m = n$ and there exists a unit v of R and a permutation $p : \{1,\ldots,m\} \to \{1,\ldots,m\}$ with $ve_iv^{-1} = f_{p(i)}$, $1 \le i \le m$.

<u>Proof</u>. Let $R \to R/\mathrm{Rad}(R) = S$ with $r \to \bar{r}$ be the natural morphism. By the next section $\sum_{i=1}^m S\bar{e}_i = \sum_{j=1}^n S\bar{f}_j$ and $m = n$. Further for a suitable permutation p, $S\bar{e}_i \simeq S\bar{f}_{p(i)}$. Then, by (VII.10), $Re_i \simeq Rf_{p(i)}$. Thus, by (VII.2), there exist v_i and w_i with $w_iv_i = e_i$ and

$v_i w_i = f_{p(i)}$. Set

$$v = \sum_{i=1}^{n} f_{p(i)} v_i e_i$$

and

$$w = \sum_{j=1}^{n} e_j w_j f_{p(j)} .$$

Then $vw = wv = 1$ and $ve_j = f_{p(j)} v$.

<u>(VII.14)</u> <u>COROLLARY</u>. Let ${}_R R = \oplus \sum_{i=1}^{m} L_i$ and ${}_R R = \oplus \sum_{j=1}^{n} N_j$ where L_i and N_j are indecomposable left ideals. Then $m = n$ and after renumbering $L_i \simeq N_i$ (as R-modules) for $1 \le i \le m$.

<u>(VII.1)</u> <u>EXERCISE</u>. If M is a free R-module of R-dimension m , use (VII.1) and show that $\mathrm{End}_R(M)$ is isomorphic to the $m \times m$ matrix ring over the opposite ring R^0 of R .

<u>(VII.2)</u> <u>EXERCISE</u>.

(a) Show that

$$\mathrm{Hom}_Z(Z/Zp^m, Z/Zp^n) \simeq Z/Zp^t$$

where $t = \min(m,n)$ and p is a prime.

(b) Show that

$$\mathrm{Hom}_Z(Z/Zp^m, Z/Zq^n) = 0$$

for p and q distinct primes.

(c) Use (a), (b) and (VII.1) and illustrate how one may determine $\mathrm{Hom}_Z(A,A)$ for a finite Abelian group A.

<u>(VII.3)</u> <u>EXERCISE</u>. Let $M = M_1 \oplus \cdots \oplus M_m$ be R-modules. If $m \geq 2$, show that $\mathrm{End}_R(M)$ is generated by $\mathrm{Aut}_R(M)$ as a Z-module.

<u>(VII.4)</u> <u>EXERCISE</u>. Let $M = M_1 \oplus \cdots \oplus M_m$ be R-modules. Suppose, for $i \neq j$, $\mathrm{Hom}_R(M_i,M_j) = 0$. Show $\mathrm{End}_R(M) \simeq R_1 \oplus \cdots \oplus R_m$ where $R_i = \mathrm{End}_R(M_i)$ for $1 \leq i \leq m$.

<u>(VII.5)</u> <u>EXERCISE</u>. Let $M = M_1 \oplus \cdots \oplus M_m$ and $N = N_1 \oplus \cdots \oplus N_n$ be R-modules.

(a) Show that

$$\mathrm{Hom}_R(M,N) \simeq \{[\phi_{ij}] \mid \phi_{ij} \in \mathrm{Hom}_R(M_j,N_i)\}$$

(as an Abelian group).

(b) Let

$$\phi = \begin{bmatrix} \phi_1 & \phi_2 \\ \phi_3 & \phi_4 \end{bmatrix} : M_1 \oplus M_2 \to N_1 \oplus N_2$$

be an R-morphism. Show that if ϕ and ϕ_1 are R-isomorphisms then $M_2 \simeq N_2$. (Hint: First consider the case where $\phi_3 = 0$.)

(c) Show that if $\text{End}_R(M_i)$ is local for $1 \leq i \leq m$, $M \simeq N$ and N_j is indecomposable for $1 \leq j \leq n$, then $m = n$ and there is a permutation σ of $\{1,2,\ldots,m\}$ such that $M_i \simeq N_{\sigma(i)}$ for $1 \leq i \leq m$.
(This is one method of proof of the Krull-Schmidt Theorem.)

(VII.6) EXERCISE. Let $C(R)$ denote the set of central idempotents of a ring R.

(a) Show that R is indecomposable (as a ring) if and only if $C(R) = \{0,1\}$.

(b) Show $C(R)$ forms a Boolean algebra under

$$e \vee f = e + f - ef$$
$$e \wedge f = ef .$$

The complement of e is $1 - e$ and $f \leq e$ if and only if $ef = f$.

(c) A non-empty subset $J \subseteq C(R)$ such that for e and f in J then $e \vee f$ is in J; and, for e in J and $f \leq e$ then f is in J, is called an ideal of $C(R)$. Show that if I is an ideal of R then $I \cap C(R)$ is an ideal of $C(R)$. Further, show that if J is an ideal of $C(R)$, then $\overline{J} = \{re \mid r \text{ in } R,\ e \text{ in } J\}$ is an ideal of R and is the smallest ideal containing J.

<u>(VII.7)</u> <u>EXERCISE</u>. Let e be an idempotent of a ring R. Show that f is an idempotent and $Re = Rf$ if and only if $f = e + (x - ex)e$ for some x in R.

<u>(VII.8)</u> <u>EXERCISE</u>. (Azumaya) Let R be a ring.

(a) Let e_1 and f_2 be two idempotents with $Re_1 \supset Rf_2$. Show there exist idempotents e_2 and f_1 with $e_1 > f_1$, $e_2 > f_2$ and $Re_1 = Re_2$, $Rf_1 = Rf_2$.

(b) Let e and f be idempotents and $e > f$. If L is a left ideal of R which is a direct summand of ${}_RR$ and $Re \supset L \supset Rf$ then show there is an idempotent g with $e > g > f$ and $L = Rg$.

<u>(VII.9)</u> <u>EXERCISE</u>. (Azumaya) Let R be a ring and e and f idempotents in R. Then e is said to be isomorphic to f, written $e \simeq f$, if $Re \simeq Rf$ (as left R-modules). Let e and f be minimal. Show the following are equivalent.

(a) $e \simeq f$

(b) f is in ReR

(c) $ReR = RfR$

(d) $eRf \not\subseteq \text{Rad}(R)$

<u>(VII.10)</u> <u>EXERCISE</u>. (Matrix Units) Let R be a ring. A set $\{e_{ij}\}$ $(i,j = 1,\dots,n)$ in R is called a set of <u>matrix</u> <u>units</u> if

$$\sum_{i=1}^{n} e_{ii} = 1$$

and $$e_{ij}e_{st} = \delta_{js}e_{it} \qquad (\delta = \text{Kronecker delta}) .$$

(a) Show that if R contains a set of matrix units $\{e_{ij}\}$ then

$$R \simeq M_n(S)$$

where $S = e_{11}Re_{11}$.

(b) Show that if ${}_RR = L_1 \oplus \cdots \oplus L_n$ (L_i left ideals) such that $L_i \simeq L_j$ (as left R-modules) for all i and j , the R contains a set of matrix units.

(This provides a slightly different approach to (VII.3).)

<u>(VII.11)</u> <u>EXERCISE</u>. (Automorphisms of $M_n(R)$) A ring R is called <u>stable</u> with respect to the ring automorphisms of $M_n(R)$ if for any pair of sets of matrix units $\{E_{ij}\}$ and $\{F_{ij}\}$ in $M_n(R)$ there is an invertible matrix P with

$$P^{-1}E_{ii}P = F_{ii} \qquad 1 \le i \le n .$$

(a) Show that if $P^{-1}E_{ii}P = F_{ii}$ for $1 \le i \le n$ then $P^{-1}E_{ij}P = F_{ij}$ for all i and j . (Hint: Use method of proof of (VII.13).)

(b) Let R be stable and $\wedge : M_n(R) \to M_n(R)$ be a ring automorphism.

(1) Show $\wedge(E_{ij}) = P^{-1}E_{ij}P$ for all i and j .

(2) Define $\Phi : M_n(R) \to M_n(R)$ by $\Phi(A) = P\,\Lambda(A)P^{-1}$
Show that, if $A = [a_{ij}]$ then $\Phi(A) = [\sigma(a_{ij})] = A^{\sigma}$
for some ring automorphism σ of R.

(3) Conclude from (1) and (2) that

$$\Lambda(A) = P^{-1}A^{\sigma}P \quad \text{for all } A \text{ in } M_n(R) .$$

(c) (1) Show that finite fields are stable.

(2) Show that matrix rings over stable rings are stable.

(3) Show that if $R/\mathrm{Rad}(R)$ is stable then R is stable.

<u>(VII.12)</u> <u>EXERCISE</u>. An element v in a ring R is an <u>involution</u> if $v^2 = 1$. If the characteristic of R is not 2^{λ} show that v is an involution if and only if $e = \frac{1}{2}(1 + v)$ is an idempotent.

<u>(VII.13)</u> <u>EXERCISE</u>. (Farahat) Let e be a non-zero idempotent in R and R^* denote the group of units of R.

(a) Show there exists a unique maximal multiplicative group $G_e(R)$ having e as its identity. If $e = 1$, $G_e(R) = R^*$.

(b) Show $G_e(R) = (eRe)^*$.

VIII. FINITE SIMPLE RINGS

This chapter - the longest in this text - examines the matrix ring over a finite field. The emphasis is on facets of ring theory rather than linear algebra. We begin in (A) by showing that a semi-simple ring is a direct sum of matrix rings over finite fields. Having this, the attention is shifted to the simple summands of the semi-simple ring, i.e., the matrix ring $M_n(k)$ over a finite field k. Section (B) describes the subfields and the simple subrings of $M_n(k)$. On the other hand, (C) examines the automorphisms and a Galois theory for $M_n(k)$ and by this we mean that a lattice inverting bijection is established between certain subrings of $M_n(k)$ and a collection of subgroups of $\mathrm{Aut}(M_n(k))$. A brief examination of the units of $M_n(k)$, i.e., the general linear group, is given in (D) with emphasis on the normal subgroups and Sylow subgroups of the group of units. Finally, (E) describes first the rings R satisfying $R/Rad(R) = M_n(k)$ and second indicates a little about the algebras which may be embedded in $M_n(k)$.

Section (A). Finite Semi-Simple Rings.

In this section we describe the structure of R when the radical of R is 0 .

(VIII.1) THEOREM. Let R have $Rad(R) = 0$; i.e., let R be semi-simple. Then ${}_RR$ is a direct sum of minimal left ideals.

Proof. We have that ${}_RR = L_1 \oplus \cdots \oplus L_n$ where L_i are indecomposable and $L_i = Re_i$. But the maximal proper submodule of L_i is $N_i = L_i \cap Rad(R) = 0$. Thus the L_i are minimal.

Second Proof. We also give a direct proof which does not utilize (VII.9). Let $M_1,\ldots,M_n$ be the maximal left ideals of R . Then $0 = Rad(R) = \cap M_i$. We may assume M_i does not contain $L_i = \bigcap_{j \neq i} M_j$ (or else we may eliminate M_i). Then $M_i + L_i = R$ and $M_i \cap L_i = 0$. Thus as left R-modules $L_i \simeq R/M_i$ and L_i is minimal.

Then $L_i = Re_i$ (by VII.6)) and $M_i = R(1 - e_i)$. Set $e = e_1 + \cdots + e_n$. Then

$$e - 1 = (e_i - 1) + \sum_{j \neq i} e_j$$

is in $M_i + M_i$ and thus in M_i. Thus $e - 1$ is in $\cap M_i$, i.e., $e = 1$. Thus $1 = e_1 + \cdots + e_n$ and ${}_RR = L_1 \oplus \cdots \oplus L_n$.

Hence, if $\mathrm{Rad}(R) = 0$, then ${}_RR = L_1 \oplus \cdots \oplus L_n$ where the L_i are minimal. The <u>homogeneous component</u> $H(L_i)$ of L_i, briefly H_i, is the direct sum of <u>all</u> minimal left ideals L with $L \simeq L_i$ (as left R-modules).

<u>(VIII.2)</u> <u>THEOREM</u>. Let R be semi-simple and suppose

$${}_RR = \oplus \sum_{i=1}^{n} L_i \qquad \text{(minimal left ideals)}$$

and

$$R_R = \oplus \sum_{j=1}^{m} M_j \qquad \text{(minimal right ideals)}.$$

Then

(a) $m = n$.

(b) ${}_RR$ and R_R have the same homogeneous components.

(c) The homogeneous components are minimal two-sided ideals.

<u>Proof</u>. Let H be a component. Then $H = \oplus \sum Rf$ where Rf is a minimal left ideal and $\sum$ extends over all idempotents f with $Rf \simeq Re$. But $Rf \simeq Re$ if and only if (VII.2) $eR \simeq fR$. Thus $\overline{H} = \oplus \sum fR$ is a component of R_R corresponding to H. We claim that $H = \overline{H}$.

Suppose $\phi : {}_R R \to {}_R R$ is an R-morphism. Then

$$\phi(H) = \phi(\oplus \sum Rf)$$
$$\subseteq H$$

since $\phi(Rf) \subseteq H$. But ϕ is a right multiplication map and thus H is an ideal.

Hence $\overline{H} = \oplus \sum fR \subset H$. Similarly $H \subset \overline{H}$. Thus $H = \overline{H}$ and $m = n$.

To show that H is minimal, let $K \neq 0$ be an ideal in H . Then ${}_R K$ is a direct sum of minimal left ideals of the form Re . But if $Rf \simeq Re$ then $f = vu = veu$ and $Rf = Reu \subset Ku \subset K$; i.e., $H \subset K$.

Thus, we easily see that

$$R = H_1 \oplus \cdots \oplus H_t$$

where the H_i are ideals. We apply (VI.1) and $H_i = Rc_i$ where c_i is a central idempotent and H_i is a ring with identity c_i .

Further,

$$H_i = Rc_i = \oplus \sum_{j=1}^{t(i)} Re_{i_j} ,$$

where

$$Re_{i_j} \text{ are minimal left ideals}$$

and

$$Re_{i_j} \simeq Re_{i_k} \ , \quad 1 \leq j \ , \quad k \leq t(i) \ .$$

It is clear that

$$c_i = c_i(e_{i_1} + \cdots + e_{i_{t(i)}}) = e_{i_1} + \cdots + e_{i_{t(i)}} \ .$$

Now apply (VII.3) to the H_i . This gives the following structure.

(VIII.3) THEOREM. Let $H = \oplus \sum Re$ be a homogeneous component of a semi-simple ring R . Then, as a ring, H is isomorphic to a complete matrix ring over the finite field eRe .

(VIII.4) THEOREM. (Structure Theorem for Finite Semi-Simple Rings) If R is a semi-simple ring then R decomposes uniquely (up to order of summands) as a finite direct sum of matrix rings over finite fields.

Proof. We need only show the uniqueness. This will follow from several lemmas.

LEMMA A. If $R \simeq M_n(k)$ (as rings) where k is a finite field, then n and k are unique.

Proof. The left $M_n(k)$-module $M_n(k)$ has a composition series

$$L_1 \oplus \cdots \oplus L_n \supset L_2 \oplus \cdots \oplus L_n \supset \cdots \supset L_{n-1} \oplus L_n \supset L_n \supset (0)$$

where if E_{ij} is the standard matrix unit then $L_i = M_n(k)E_{ii}$. This series has length n. Thus since $R \simeq M_n(k)$, ${}_R R$ has a composition series of length n. By the Jordan-Hölder Theorem the length of a composition series is unique. Thus n is unique.

Since $|k| = p^m$ for some prime p and integer m, $|R| = (p^m)^{n^2}$. This gives the uniqueness of k.

The next lemma is a standard exercise.

LEMMA B. Let k be a finite field. Then $M_n(k)$ is a simple ring.

The following is straightforward.

LEMMA C. Let $R = \oplus \sum_{j=1}^{n} R_j$ where R_j is a ring with identity 1_j. Then each left ideal L of R has the form

$$L = \oplus \sum_{j=1}^{n} L_j$$

where L_j is a left ideal of R_j. Conversely, each such sum is a left ideal of R.

Lemma C gives the next result.

LEMMA D. A finite direct sum of simple rings is a semi-simple ring.

We now prove (VIII.4): Suppose that R is semi-simple and

$$R = \oplus \sum_{i=1}^{n} R_i$$
$$= \oplus \sum_{j=1}^{m} S_j \ . \qquad R_i, S_j \text{ simple rings.}$$

Then S_j is an ideal of R and by (C)

$$S_j = \oplus \sum \overline{Y}_i$$

where $\overline{Y}_i = R_i$ or $\overline{Y}_i = (0)$. But S_j is simple thus for some index i $S_j = \overline{Y}_i = R_i$ and $\overline{Y}_k = (0)$ $k \neq i$. Thus $m = n$ and after renumbering $R_i = S_i$. Now apply Lemma A.

Conversely, any finite sum of matrix rings over finite fields is a semi-simple ring.

We conclude this section with some properties of semi-simple and simple rings.

<u>(VIII.5)</u> <u>THEOREM</u>. The following are equivalent:

(a) The ring R is semi-simple.

(b) Every left ideal of R is principal and a direct summand of ${}_RR$.

(c) If a is in R then there is an a' with $aa'a = a$.

<u>Proof</u>. By examining the left ideals of $M_n(k)$ it is easy to see that (a) implies (b). Assume (b) and let a be in R. Then Ra is a direct summand and there is an idempotent e with $Ra = Re$. Then

$e = a'a$ and $aa'a = ae = a$. To show (c) implies (a) let $a \neq 0$ be in R . Then $1 - aa'$ is a zero divisor. Hence a is not in the radical of R and $Rad(R) = 0$.

<u>(VIII.6)</u> <u>THEOREM</u>. (Schafer). A simple ring is generated by $n \leq 2$ elements.

<u>Proof</u>. (Schafer [97]). If R is commutative, then R is a finite field. Thus the units R^* of R form a cyclic multiplicative group. A generator of R^* generates R as a ring.

If R is not commutative then we may assume $R = M_n(k)$ for some n and finite field k . Let

$$y = \begin{bmatrix} 0 & 1 & 0 & \cdots & 0 \\ 0 & 0 & 1 & \cdots & 0 \\ \cdot & & & & \\ \cdot & & & & \\ \cdot & & & & \\ 0 & 0 & 0 & \cdots & 1 \\ 1 & 0 & 0 & \cdots & 0 \end{bmatrix} .$$

Then $y^n = 1$. If E_{ij} denotes the standard matrix unit then

$$y^{n-i+1}E_{11} = E_{i1}$$

and

$$E_{11}y^{j-1} = E_{1j} .$$

Hence, $E_{ij} = E_{i1}E_{1j} = y^{n-i+1}E_{11}y^{j-1}$.

For a in R , $a = [\alpha_{ij}]$,

$$a = \sum \alpha_{ij}E_{ij} = \sum y^{n-i+1}(\alpha_{ij}E_{11})y^{j-1} .$$

If Θ generates k* then $\alpha_{ij} = \Theta^{S_{ij}}$ and $\alpha_{ij}E_{11} = (\Theta E_{11})^{S_{ij}}$. Thus y and ΘE_{11} generate R .

<u>(VIII.7)</u> <u>THEOREM</u>. Let R be a simple ring and M a finitely generated R-module. Then $M = M_1 \oplus \cdots \oplus M_m$ where $M_i \simeq L$ (as R-module) for $1 \leq i \leq n$ where L is a minimal left ideal of R .

<u>Proof</u>. The module M is a direct sum of indecomposable R-modules. Thus it suffices to assume M is indecomposable and show $M \simeq L$ where L is a minimal left ideal of R . Note any two minimal left ideals of R are isomorphic as R-modules. Let ${}_RR = Re_1 \oplus \cdots \oplus Re_n$ be a decomposition of R into minimal left ideals. Then

$$M = \sum_{m\in M} \sum_{i=1}^{n} Re_iM .$$

The map $Re_i \to Re_im$ by $re_i \to re_im$ is an R-morphism and since Re_i is a minimal left ideal it has kernel either 0 or Re_i . Thus M is isomorphic to a sum of minimal left ideals of R . It is easy to see that such a sum is direct (hence M is decomposable) unless $M \simeq Re_im$ for some m in M and idempotent e_i .

Section (B). Simple Subrings of $M_n(k)$.

This section concerns the simple subrings in $M_n(k)$, i.e., matrix rings in $M_n(k)$.

(VIII.8) THEOREM. (Distinguished Basis Theorem). Let k be a finite field and V be a two-sided k-vector space of left dimension m . Let k_1 be the maximal subfield of k satisfying $\alpha v = v\alpha$ for all α in k_1 and v in V . Then there exists a left k-basis $\{b_1,\ldots,b_m\}$ of V and k_1-automorphisms $\{\sigma_1,\ldots,\sigma_m\}$ of k such that

$$\alpha b_i = b_i \sigma_i(\alpha)$$

for all α in k .

Proof. We have $k \simeq k_1[X]/(f)$ where f is an irreducible polynomial of degree $r = [k:k_1]$. In $k[X]$, f splits into distinct linear factors $f(X) = (X - a_1) \cdots (X - a_r)$ for a_i in k and $a_i \neq a_j$. Define $f_i(X) = \prod_{j\neq i} (X - a_j)$. Then $f_i(a_i) \neq 0$ and

$$(*) \qquad \textstyle\sum_{i=1}^{r} [f_i(a_i)]^{-1} f_i(X) = 1 .$$

Let $\Phi : V \to V$ be the k-linear left morphism given by $\Phi(v) = va_1$. Observe $f(\Phi)(v) = vf(a_1) = 0$. Applying (*) we have the identity morphism

$$i_V = \sum [f_i(a_i)]^{-1} f_i(\Phi) : V \to V$$

which induces the natural decomposition

$$V = V_1 + \cdots + V_r$$

where $V_i = [f_i(a_i)]^{-1} f_i(\Phi)(V)$. It is easy to see that the sum $V = V_1 + \cdots + V_r$ is direct.

Let v be in V_i . Then for some w in V , $v = [f_i(a_i)]^{-1} f_i(\Phi)(w)$ and

$$\begin{aligned}(\Phi - a_i)(v) &= [f_i(a_i)]^{-1} f(\Phi)(w) \\ &= 0 .\end{aligned}$$

Thus $\Phi(v) = a_i v$. By the definition of Φ , $\Phi(v) = va_1$. That is, for all v in V_i , $a_i v = va_1$. Since a_1 and a_i are zeros of f in $k[X]$, there exists a k_1-automorphism σ_i with $\sigma_i(a_1) = a_i$. If α is in $k = k_1[a_1]$ then $\alpha = c_0 + c_1 a_1 + \cdots + c_{r-1} a_1^{r-1}$ for c_i in k_1 . It is easy to see that $v\alpha = \sigma_i(\alpha)v$ for all v in V . The result follows.

Let A and B be algebras[1] over a finite field with bases $\{a_1, \ldots, a_m\}$ and $\{b_1, \ldots, b_n\}$ respectively. Define the tensor product (direct product) $A \otimes B$ to be an algebra with basis $\{a_i b_j\}_{i=1,j=1}^{m,n}$ (or more appropriately $a_i \otimes b_i$) with multiplication given by $(a_i b_j)(a_s b_t) = (a_i a_s) \otimes (b_j b_t)$ and extended by the distributive law.

---

[1] A is an algebra over k if A is a ring and A possesses a k-space structure which is compatible with the ring structure in the obvious fashion.

EXAMPLES

(1) Let A be an algebra over a finite field k and K an extension field of k. Then $K \otimes A$, written A_K, is an algebra over K with the same basis as A and the same multiplication between basis elements as in A. (This is called extension of scalars.)

(2) Let A be an algebra over a finite field k and $B = M_n(k)$. Then $A \otimes B \simeq M_n(A)$.

(3) For a finite field k, $M_n(k) \otimes M_m(k) \simeq M_{mn}(k)$.

(VIII.9) THEOREM. Let $k_1 = GF(p^t)$ and $k_2 = GF(p^s)$ be finite fields (p a prime). If $d = \text{g.c.d.}(s,t)$ and $l = \text{l.c.m.}(s,t)$ then

(a) $k_1 \cap k_2 = GF(p^d)$

(b) $k_1 \otimes k_2 = GF(p^l)$ where k_1 and k_2 are regarded as algebras over $k_1 \cap k_2$.

Proof. In the algebraic closure of $GF(p)$ consider the set $F = \{\sum a_i b_i \mid a_i \text{ in } k_1, b_i \text{ in } k_2\}$. Clearly $F \subset GF(p^l)$ is a finite field of p^v elements. Since k_1 and k_2 are in F, t divides v and s divides v. Hence $v = l$ and $F = GF(p^l)$.

Define $F \to k_1 \otimes k_2$ by $\sum a_i b_i \to \sum a_i \otimes b_i$ (tensor product over $GF(p^d)$). It is easy to see that this is an injective field morphism. Since $|F| = |k_1 \otimes k_2|$, it is an isomorphism.

We fix for the remainder of this section the following notation and conventions:

(a) $k = GF(p^t)$ (p a prime).

(b) V = vector space over k of dimension n.

(c) $k_0 = Z/Zp$ (prime field).

(d) $L = M_n(k)$ (we often identify L with $End_k(V)$).

(e) For R a subring of L,

$C_L(R) = \{\sigma \text{ in } L \mid \alpha\sigma = \sigma\alpha \text{ for all } \alpha \text{ in } R\}$ = <u>centralizer</u> of R in L.

(f) If k_1 is any finite field with $k_0 \subseteq k_1 \subseteq k$ then k_1 is identified with its corresponding field of scalar matrices in L.

(g) If R is a subring of L, it is assumed the identity of R is the same as the identity of L.

<u>(VIII.10)</u> <u>THEOREM</u>. Let K be a finite field containing k. If $[K:k] = m$, then

(a) $k \subseteq K \subseteq L$ if and only if for some integer q then $mq = n$, i.e., m divides n.

(b) If m divides n, then $C_L(K) = M_q(K)$.

<u>Proof</u>. Suppose K is a finite field in L with $k \subseteq K$. Then V is a K-space under $\alpha v = \alpha(v)$ for α in K and v in V. Let $q = \dim_K(V)$ and $\{v_j\}$ be a K-basis of V. Then if $\{\alpha_i\}$ is a k-basis of K, $\{\alpha_i v_j\}$ is a k-basis of V. Thus

$$n = \dim_k(V) = \dim_K(V)\dim_k(K) = qm .$$

Conversely, let α be an m by m matrix over k $(mq = n)$ with irreducible characteristic polynomial c_α. Let $\beta = \mathrm{diag}[\alpha \oplus \cdots \oplus \alpha]$ (q blocks) in L. Then the minimal polynomial m_β of β in L is c_α and in L

$$k[\beta] \simeq k[X]/m_\beta = k[X]/c_\alpha \simeq K .$$

Finally, $C_L(K) = \{\alpha \text{ in } L \mid \alpha\sigma = \sigma\alpha \text{ for all } \sigma \text{ in } K\} = \mathrm{End}_K(V)$. But V as a K-space has dimension q. Thus $C_L(K) \simeq M_q(K)$

(VIII.11) THEOREM. Let S be a simple ring with $k \subseteq S \subseteq L$. Then, clearly

$$S = M_v(K)$$

where K is a finite field, $k \subseteq K$, $[K:k] = m$ and $mq = n$. Further, if $\overline{S} = C_L(S)$ then

(a) $\overline{S} = M_w(K)$

(b) $S = C_L(\overline{S})$

(c) There is the following lattice diagram:

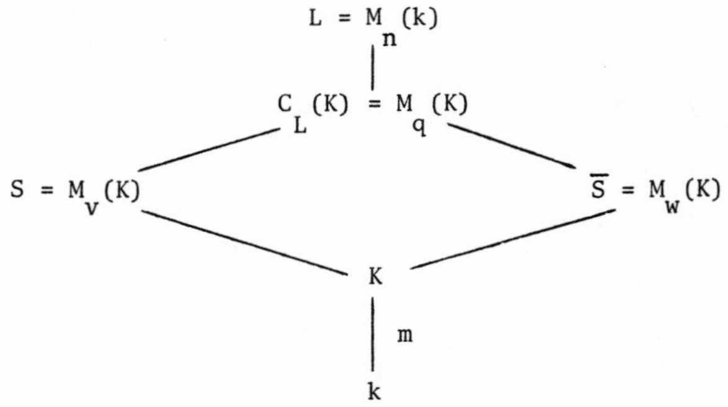

with $\dim_k(S) = mv^2$, $\dim_k(\overline{S}) = mw^2$ and $(mv^2)(mw^2) = m^2q^2$.

Proof. Let S be a simple ring with $k \subseteq S \subseteq L$. Then $S = M_v(K)$ for some v and some finite field K with $k \subseteq K \subseteq L$.

Let $m = [K:k]$. Then $n = mq$. Further by the previous result $C_L(K)$ is simple and $C_L(K) = M_q(K)$. Let $T = C_L(K)$.

Then $S \subseteq T$ and $C_L(S) \subseteq C_L(K)$ since $K \subseteq S$. Let W be a K-vector space of dimension q and regard $T = C_L(K) = M_q(K)$ as $\mathrm{End}_K(W)$. Then W is naturally an S-module. By (VIII.7) $W = W_1 \oplus \cdots \oplus W_t$ as an S-module where $W_i \simeq L$ (L a minimal left ideal of S) .

But $C_T(S) = \mathrm{End}_S(W)$ and by (VII.3) $\mathrm{End}_S(W) \simeq \mathrm{End}_S(L \oplus \cdots \oplus L) \simeq M_t(K)$. To determine t remember L is a minimal left ideal of S if and only if $L = Se$ where e is a minimal idempotent. Since

$S = M_v(K)$ by (VII.13) there is a similarity transformation which carries e into the matrix unit E_{11}. Thus take $L = SE_{11} = \{\langle a_1, \ldots, a_v\rangle^t \mid a_i$ in $K\}$. Thus L is a K-space of dimension v. Then $q = \dim(W) = tv$.

We conclude this section with the Noether-Skolem Theorem for L.

(VIII.12) THEOREM. (Noether-Skolem). Let S and T be simple subrings of L containing k. If $\phi : T \to S$ is a k-isomorphism, then there exists an invertible ρ in L with

$$\phi(\alpha) = \rho\alpha\rho^{-1}$$

for all α in T. (See Exercises (VIII.26) and (VIII.27)).

Proof. Regard L as $\mathrm{End}_k(V)$. Then V is naturally a left T-module in two ways

$$t \cdot v = t(v) \qquad \text{(denoted } {}_TV\text{)}$$

and

$$t \cdot v = \phi(t)(v) \qquad \text{(denoted } {}_SV\text{)}.$$

As in the last proof, each T-module is a direct sum of isomorphic left ideals of T. By counting dimensions we have the same number of summands. Thus as T-modules, ${}_TV \cong {}_SV$. Let the isomorphism ${}_TV \to {}_SV$ be denoted by ρ. Then for all t in T, $\rho(t \cdot v) = t \cdot \rho(v) = \phi(t)\rho(v)$. In particular, if t is in k then ρ is seen to be k-linear, i.e., ρ is in L . As mappings, if t is in T then $\rho t = \phi(t)\rho$. That is, $\phi(t) = \rho t \rho^{-1}$.

(VIII.13) COROLLARY. Let $\phi : L \to L$ be a k-algebra isomorphism. Then there is an invertible ρ in L with $\phi(\alpha) = \rho\alpha\rho^{-1}$ for all α in L .

Proof. Take L = T = S in (VIII.12).

Section (C). The Galois Theory of $M_n(k)$.

In this section we describe a Galois theory for a simple ring. By a "Galois theory" we mean that we will establish a lattice inverting bijection between a class of subrings of a simple ring and a collection of subgroups of the group of ring automorphisms of the simple ring.

The following notation and conventions are to be assumed throughout this section.

$k = GF(p^t)$ (p a prime)

V = vector space over k of dimension n

$k_0 = Z/Zp$ (prime field)

$L = M_n(k)$

A = group of all ring automorphisms of $L = M_n(k)$.

Further, for a subring R in L ,

$\mathrm{Aut}_A(R) = \{\Lambda \text{ in } A \mid \Lambda(\alpha) = \alpha \text{ for all } \alpha \text{ in } R\}$

= <u>group of automorphisms fixing</u> R

$C_L(R) = \{\sigma \text{ in } L \mid \alpha\sigma = \sigma\alpha \text{ for all } \alpha \text{ in } R\}$

= <u>centralizer of</u> R <u>in</u> L

$Z(R) = \{\sigma \text{ in } R \mid \alpha\sigma = \sigma\alpha \text{ for all } \alpha \text{ in } R\}$

= <u>center of</u> R .

For a subgroup G of A ,

$\mathrm{Rng}_L(G) = \{\sigma \text{ in } L \mid \Lambda(\sigma) = \sigma \text{ for all } \Lambda \text{ in } G\}$

= <u>fixed ring under</u> G .

Finally,

(1) If k_1 is any finite field with $k_0 \subseteq k_1 \subseteq k$ then k_1 is identified with its corresponding field of scalar matrices in L .

(2) If R is a subring of L then it is assumed the identity of R is the same as the identity of L .

We consider the description of a single ring automorphism of L .

A mapping $\rho : V \to V$ is called a <u>semi-linear transformation</u> if there exists a field automorphism $\sigma : k \to k$ such that

(1) $\rho(v + w) = \rho(v) + \rho(w)$

(2) $\rho(\alpha v) = \sigma(\alpha)\rho(v)$

for v and w in V and α in k .

Let $\{v_1,\ldots,v_n\}$ be a k-basis for V and $\rho : V \to V$ semi-linear. Then, if

$$\rho(v_i) = \sum_{j=1}^{n} a_{ji}v_j ,$$

the matrix

$$P = [a_{ji}]$$

is called the <u>matrix of</u> ρ, $\mathrm{Mat}(\rho)$, <u>relative to</u> $\{v_i\}$. We identify ρ with the pair $\langle P,\sigma\rangle$. If $\sum \alpha_i v_i$ is in V then

$$\begin{aligned}\rho(\textstyle\sum_i \alpha_i v_i) &= \textstyle\sum_i \sigma(\alpha_i)\rho(v_i)\\ &= \textstyle\sum_i \sigma(\alpha_i)(\sum_j a_{ji}v_j)\\ &= \textstyle\sum_j (\sum_i a_{ji}\sigma(\alpha_i))v_j .\end{aligned}$$

Thus, in the language of matrices,

$$\langle P,\sigma\rangle \begin{bmatrix}\alpha_1\\ \vdots\\ \alpha_n\end{bmatrix} = P \begin{bmatrix}\sigma(\alpha_1)\\ \vdots\\ \sigma(\alpha_n)\end{bmatrix} .$$

If $\{w_1,\ldots,w_n\}$ is a second basis and

$$w_i = \sum_j \beta_{ji}v_j$$

then

$$\mathrm{Mat}(\rho) = B^{-1}PB^{\sigma}$$

is the matrix of ρ relative to $\{w_1,\ldots,w_n\}$ where $B = [\beta_{ij}]$ and $B^{\sigma} = [\sigma(\beta_{ij})]$.

(VIII.14) THEOREM. Let $\Lambda : L \to L$ be a ring automorphism. Then there exists an invertible semi-linear transformation $\rho : V \to V$ such that

$$\Lambda(\alpha) = \rho\alpha\rho^{-1}$$

for all α in L .

Proof. Exercise. (See exercises in this section.)

Denote Λ above by Λ_ρ . Observe that $\Lambda_\rho = \Lambda_{\alpha\rho}$ for any $\alpha \neq 0$ in k . If ρ and ϕ are semi-linear such that $\rho \leftrightarrow \langle P,\sigma\rangle$ and $\phi \leftrightarrow \langle Q,\tau\rangle$ then $\rho\phi \leftrightarrow \langle PQ^\sigma, \sigma\tau\rangle$.

Suppose $\Lambda_\rho = \Lambda_\phi$. Then for all α in L

$$P\alpha^\sigma P^{-1} = Q\alpha^\tau Q^{-1} .$$

Equivalently,

$$\alpha^\sigma (P^{-1}Q) = (P^{-1}Q)\alpha^\tau .$$

If $\alpha = E_{ij}$ (elementary matrix) then

$$\alpha^\sigma = E_{ij}^\sigma = E_{ij} = E_{ij}^\tau = \alpha^\tau .$$

That is, $E_{ij}(P^{-1}Q) = (P^{-1}Q)E_{ij}$ for all i and j . Thus $P^{-1}Q$ is a scalar matrix, $P^{-1}Q = \delta$ for some δ in k . Hence $Q = \delta P$ and $\sigma = \tau$. We conclude that $\Lambda_\rho = \Lambda_\phi$ if and only if $Q = \delta P$ ($\delta \neq 0$ in k) and $\sigma = \tau$.

The next result is basic to the Galois theory.

(VIII.15) THEOREM. (Hochschild). Let G be a subgroup of $A = \mathrm{Aut}(L)$ such that the only inner-automorphism in G is the identity. Let

$$S = \mathrm{Rng}_L(G)$$
$$k_1 = S \cap k .$$

Then

(a) S is a simple k_1-algebra (indeed $S \simeq M_v(k_1)$ for some v).

(b) $L = Sk$ where Sk is the k_1-algebra generated by all finite sums $\sum_i s_i\alpha_i$ with s_i in S and α_i in k. Indeed, $Sk \simeq S \otimes_{k_1} k$.

Proof. (Hochschild, [34]). We will imbed S in a simple algebra with center k_1 and use (VIII.11). The simple algebra is chosen to be the semi-linear group ring (L,G) of L with respect to G. The ring (L,G) is defined as follows:

(L,G) is a free L-module with free L-basis $\{u_1 = 1, u_\sigma, \ldots, u_\tau\}$ where $G = \{1, \sigma, \ldots, \tau\}$.

Then multiplication is given by

$$\alpha u_\sigma = u_\sigma \sigma(\alpha)$$
$$u_\sigma u_\tau = u_{\sigma\tau}$$

for α in L (and extended linearly).

Imbed L in (L,G) by

$$\rho \to \rho u_1 = \rho 1$$

and take $\rho 1 \equiv \rho$. Let k_1 be the center of (L,G) .

The first step is to show (L,G) is simple. Suppose J is a two-sided non-zero ideal in (L,G) and

$$0 \neq \sum_{i=1}^{r} \alpha_i u_{\sigma_i} \qquad (\alpha_i \text{ in } L)$$

is in J . Further select this non-zero element such that r is minimal. Then $\sigma_i \neq \sigma_j$, $1 \leq i , j \leq r$ and $i \neq j$.

First suppose r = 1 . Then for α in L with $\alpha \neq 0$, αu_σ is in J . Thus

$$(\alpha_1 \alpha \alpha_2) u_\sigma = \alpha_1 (\alpha u_\sigma) \sigma^{-1}(\alpha_2)$$

is in J . Hence since $L\alpha L = L$ we have u_σ in J . Thus $1 = u_{\sigma^{-1}} u_\sigma$ is in J and J = (L,G) .

Now suppose r > 1 . By the above we may assume that $\alpha_r = 1$. We have

$$(*) \qquad \alpha_1 u_{\sigma_1} + \ldots + \alpha_{r-1} u_{\sigma_{r-1}} + u_{\sigma_r} \neq 0$$

in J . Multiply (*) on the right by α . Then multiply (*) on the left by $\sigma_r^{-1}(\alpha)$. Subtract the two expressions and obtain

$$\sum_{i=1}^{r-1} (\alpha \alpha_i - \alpha_i \sigma_i \sigma_r^{-1}(\alpha)) u_{\sigma_i}$$

in J. By the minimality of r this element is zero and since the $\{u_\sigma\}$ are L-free,

(**) $$\alpha\alpha_i = \alpha_i\sigma_i\sigma_r^{-1}(\alpha) .$$

Thus, $L\alpha_i = \alpha_i L$ and $L\alpha_i$ is a two-sided ideal. Since L is simple, $L\alpha_i = L$. Thus, there exists an element α_i^{-1} in L with $\alpha_i^{-1}\alpha_i = 1$. Then (**) implies

$$\alpha_i^{-1}\alpha\alpha_i = \sigma_i\sigma_r^{-1}(\alpha) .$$

That is, $\sigma_i\sigma_r^{-1}$ is an inner-automorphism. By the hypothesis on G, $\sigma_i\sigma_r^{-1}$ = identity. Thus $\sigma_i = \sigma_r$ contradicting $r > 1$.

Hence (L,G) is a simple ring.

We show next $L = C_{(L,G)}(k)$. Let $\beta = \sum_{\sigma \in G} \beta_\sigma u_\sigma$ commute with every α in k. Then $\alpha - \sigma(\alpha) = 0$, i.e., $\sigma(\alpha) = \alpha$ for all α in k. Thus σ is an inner-automorphism and hence σ is the identity. Thus $\beta = \beta_1 u_1$ and β is in L.

In a similar manner, it is easy to show the semi-linear group subring (k,G) generated by $\{u_\sigma\}$ and k in $(L.G)$ is a simple ring.

Consider the following lattice to be described below:

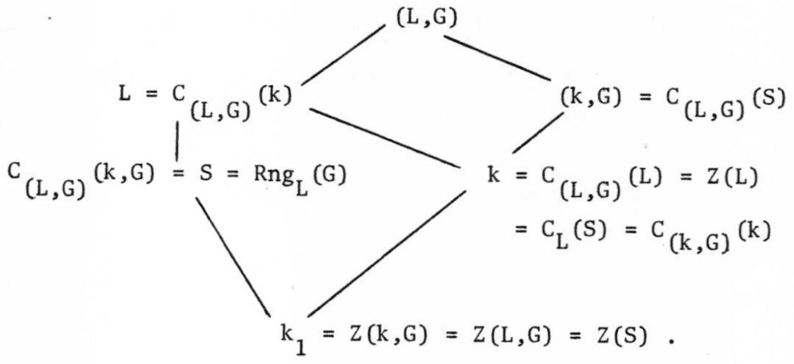

To describe the above diagram let $k_1 = Z(k,G)$.

Since $Z(L,G)$ is in L , we have $Z(L,G)$ contained in $k = Z(L)$. Thus $Z(L,G) = Z(k,G) = k_1$.

Note that $[(k,G):k] = |G| = [k:k_1]$. The last equality follows from the observation that if $\Lambda_\phi|_k = \Lambda_\Theta|_k$ where $\Phi \leftrightarrow \langle P,\sigma \rangle$ and $\Theta \leftrightarrow \langle Q,\tau \rangle$, then

$$\alpha^\sigma = \Lambda_\phi(\alpha) = \Lambda_\Theta(\alpha) = \alpha^\tau$$

for every α in k . Thus $\sigma = \tau$ and $\Lambda_\phi \Lambda_\Theta^{-1}$ = inner-automorphism . Hence $\Lambda_\Phi = \Lambda_\Theta$. Thus distinct elements in G induce distinct automorphisms of k fixing k_1 . In fact, since k_1 is the fixed subfield of k under G , G gives all the k_1-automorphisms of k , i.e., $|G| = [k:k_1]$. In particular, G is cyclic and isomorphic to $\mathrm{Aut}_k(k_1)$.

Note further $C_{(L,G)}(k,G)$ is contained in L. Thus, $S = \mathrm{Rng}_L(G) = C_{(L,G)}(k,G)$ is in L. Thus, $(k,G) = C_{(L,G)}(S)$ by (VIII.11).

Thus, $C_L(S) = (k,G) \cap L = k$.

Finally, $Z(S) = S \cap k = k_1$.

Thus, if $[k:k_1] = m = |G|$, our diagram appears as

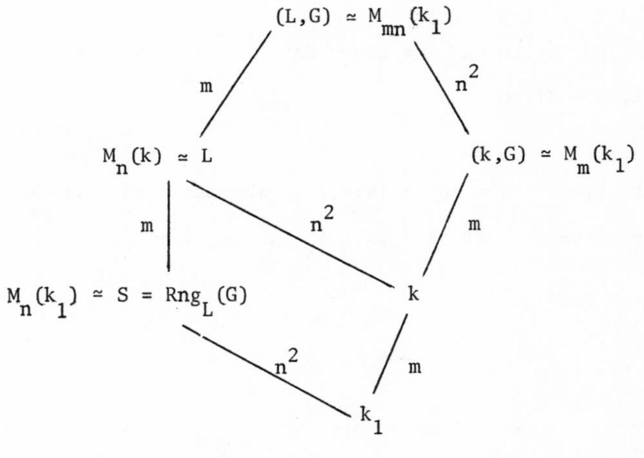

Finally, $S \otimes_{k_1} k$ is simple. Thus the natural surjective morphism

$$S \otimes_{k_1} k \to Sk$$

is an isomorphism. Hence

$$\dim_{k_1}(Sk) = \dim_{k_1}(S \otimes k) = \dim_{k_1}(S)\dim_{k_1}(k) = n^2 m .$$

But $\dim_{k_1}(L) = n^2 m$. Thus $L = Sk$.

Let G be a subgroup of $A = \text{Aut}(L)$. If ρ in L is invertible, let I_ρ denote the inner-automorphism effected by ρ in A. Let $R[G]$ denote the subring of L generated by all ρ such that I_ρ is in G. Let G_0 denote the normal subgroup of G consisting of all inner-automorphisms in G.

Recall $L = M_n(k)$ where $k = GF(p^t)$. A subgroup G of A is called <u>regular</u> if

(1) Whenever ρ in $R[G]$ is invertible then I_ρ is in G.

(2) $R[G]$ is a simple ring.

Note that k is in $R[G]$ since if $\alpha \neq 0$ is in k then $id_G = I_\alpha$. Further, if G is regular and Λ is in G then $\Lambda : R[G] \to R[G]$. To show this suppose that ρ is a generator of $R[G]$. Then $\Lambda(\rho)$ is in $R[G]$ since by direct computation the automorphism $I_{\Lambda(\rho)} = \Lambda I_\rho \Lambda^{-1}$ and thus is in G.

If S is a subring of L and an algebra over a subfield k_1 of k, denote by $[L:S]$ the rational $[L:k_1]/[S:k_1]$. Observe this does not depend on k_1.

The Galois correspondence will follow from the next two lemmas.

<u>(VIII.16)</u> <u>LEMMA</u>. Let G be a regular subgroup of $A = \text{Aut}(L)$. Then

(a) $S = \text{Rng}_L(G)$ is simple

(b) $\mathrm{Aut}_A(\mathrm{Rng}_L(G)) = G$

(c) $|G/G_0|[R[G]:k] = [L:S]$

<u>Proof</u>. (Hochschild, [34]). Since G is regular, $R[G]$ is simple. Further, k is in $Z(R[G])$. Then $C_L(R[G])$ is simple, contains k , and if $T = C_L(R[G])$,

$$[T:k][R[G]:k] = [L:k] .$$

Observe $\mathrm{Rng}_L(G) \subseteq T$.

We have noted G carries $R[G]$ into $R[G]$. Since automorphisms preserve centralizers G carries T into T . We restrict G to T . If Λ is in G and $\Lambda|_T$ is inner then $\Lambda|_T$ fixes $Z(T) = k_1$. Since $k \subseteq Z(T) = k_1$, $\Lambda|_T$ fixes k . Hence Λ fixes k and Λ is hence inner. Thus Λ is in G_0 and for some invertible ρ in L , $\Lambda = I_\rho$. Since $R[G]$ is regular, ρ is in $R[G]$. Thus ρ centralizes $T = C_L(R[G])$. That is, $\Lambda|_T = I_\rho|_T$ = identity . We conclude $G|_T \simeq G/G_0$. Thus we have a simple ring T with center k_1 and a group of automorphisms $G|_T$ acting on T and having no non-trivial inner-automorphisms. Hence we are in the setting of (VIII.15).

Let $S = \mathrm{Rng}_T(G|_T)$. Observe $S = \mathrm{Rng}_L(G)$ and S is a simple subring of T . We have the following diagram which is described more fully below:

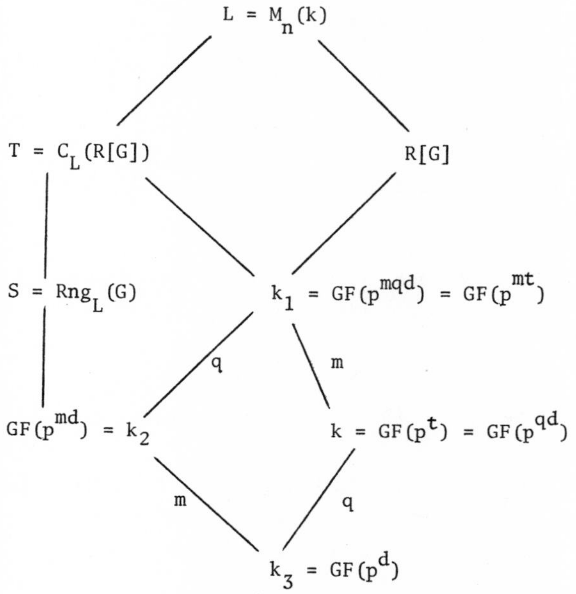

NOTE

(1) m divides n

(2) $(q,m) = 1$.

Let $k_1 = Z(T)$, $k_2 = Z(S) = S \cap k_1$ and $k_3 = S \cap k$. Observe $k_2k \subsetneq k_1$ and G maps k_2k onto k_2k . Note k_3 is the subfield of k fixed under G , and $k_2 = k_2k_3$ is the subfield of k_2k fixed under G . But $G|_{k_2k} \cong G/G_0$. Thus $[k_2k:k_2] = |G/G_0| = [k_1:k_2]$. Consequently, $k_1 = k_2k$.

Similarly $[k:k_3] = [k_1:k_2]$. Thus, since $k_3 = k_2 \cap k$ and $k_1 = k_2k$, by (VIII.9) if $k_1 = GF(p^s)$, $k = GF(p^t)$, $k_2 = GF(p^r)$ and $k_3 = GF(p^d)$ then

$$s = \text{l.c.m. } \{t,r\} \ , \quad d = \text{g.c.d. } \{t,r\} \ .$$

Assume $[k_1:k] = m$ and $|G/G_0| = q$. Then $t = md$, $r = qd$, $s = mqd$. In particular, $(m,q) = 1$.

We now compute $|G/G_0|[R[G]:k]$. By the above

$$|G/G_0|[R[G]:k] = [k:k_3]([L:k]/[T:k]) = [L:k_3]/[T:k] \ .$$

By (VIII.15), $T \simeq S \otimes_{k_2} k_1$. Thus $[T:k_1] = [S:k_2]$. Hence

$$\begin{aligned} [T:k] &= [T:k_1][k_1:k] \\ &= [S:k_2][k_2:k_3] \\ &= [S:k_3] \end{aligned}$$

and

$$|G/G_0|[R[G]:k] = [L:k_3]/[S:k_3] = [L:S] \ .$$

To finish the proof we show

$$\text{Aut}_A(\text{Rng}_L(G)) = G \ .$$

Let $G_1 = \text{Aut}_A(\text{Rng}_L(G))$. Observe $G \subseteq G_1$ and $S = \text{Rng}_L(G) = \text{Rng}_L(G_1)$.

Thus $R[G_1] \subseteq C_L(S) = R'$. But $C_L(S) = C_L(Sk)$ and a direct computation shows that $Sk \simeq S \otimes_{k_3} k$ is a simple k-algebra with center k_1 . Thus $R' = C_L(S)$ is simple. Thus R' is generated by invertible elements. But if ρ in $R' = C_L(S)$ is invertible then I_ρ is in $G_1 = \mathrm{Aut}_A(S)$ and thus ρ is in $R[G_1]$. Consequently, $R[G_1] = C_L(S)$.

Note $G_1/G_{1_0} \simeq \mathrm{Aut}_k(k_3) \simeq G/G_0$.

Thus, by the equality of (c),

$$[R[G_1]:k] = [R[G]:k] .$$

Since $R[G] \subseteq R[G_1]$, this implies $R[G] = R[G_1]$. Thus $G_{1_0} = G_0$. Since $G \subset G_1$ and $|G/G_0| = |G_1/G_0|$, $G = G_1$.

<u>(VIII.17)</u> <u>LEMMA</u>. Let R be a simple subring of L . Then

(a) $\mathrm{Aut}_A(R) = G$ is regular

(b) $\mathrm{Rng}_L(\mathrm{Aut}_A(R)) = R$.

<u>Proof</u>. Let the center of R , $Z(R) = k_1$. Then k_1 is an extension of Z/Zp . Since $k = Z(L)$, $R \cap k = k_2 \subseteq k_1$. Clearly $k_1 \cap k \subseteq k_2$. But $k_2 \subseteq k_1$ and $k_2 \subseteq k$ thus $k_2 \subseteq k_1 \cap k$. Hence $k_2 = k_1 \cap k$. Thus if $k_1 = GF(p^s)$ and $k = GF(p^t)$ then $k_2 = GF(p^d)$, $d = \text{g.c.d. } \{s,t\}$.

Hence $R \otimes_{k_2} k \simeq M_w(k_1) \otimes_{k_2} k$ (for some w)

$$\simeq (M_w(k_2) \otimes_{k_2} k_1) \otimes_{k_2} k$$

$$\simeq M_w(k_2) \otimes_{k_2} (k_1 \otimes_{k_2} k)$$

$$\simeq M_w(k_2) \otimes_{k_2} k_3$$

$$\simeq M_w(k_3)$$

where $k_3 = GF(p^q)$, $q = \text{l.c.m.}\ \{s,t\}$.

In particular, $R \otimes_{k_2} k$ is simple. Since there is a natural surjective morphism $R \otimes_{k_2} k \to Rk$, Rk is simple.

Note that $R[G] \subset C_L(R)$. But, as above, $C_L(R) = C_L(Rk)$ and Rk is a simple k-algebra. Hence $C_L(R)$ is simple and consequently generated by its units. But for each unit ρ , I_ρ is in G and hence ρ is in $R[G]$. Therefore $R[G] = C_L(R)$ is simple. Hence G is regular.

Thus G/G_0 is isomorphic to a group of automorphisms of k with fixed field, say K . By the previous proof we have the following lattice.

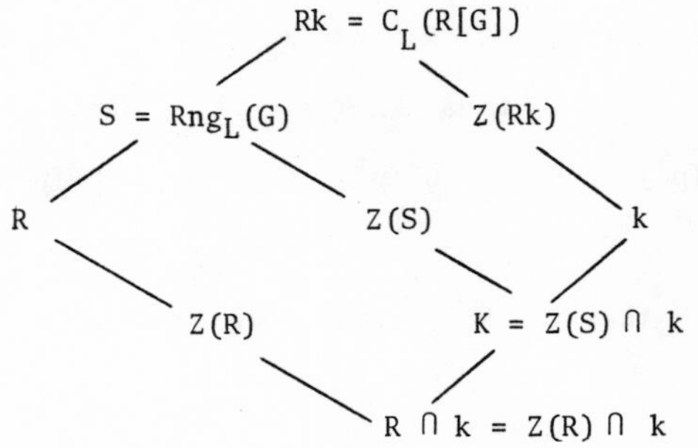

.

If $R \cap k \neq K$, then since k is a Galois extension of $R \cap k$ there is an automorphism σ of k fixing $R \cap k$ and satisfying $\sigma(\alpha) \neq \alpha$ for some α in K. Then taking any invertible ρ in $R[G]$ the semi-linear morphism $\langle \rho, \sigma \rangle = \Phi$ determines an automorphism Λ_Φ which fixes R but not S. This is not possible. Thus $R \cap k = K$. Then k_1k is a subfield of $Z(Rk)$ and, further, $Rk \simeq R \otimes_{k_1} k_1k$. Thus $k_1k = Z(Rk)$. This implies $k_1 = Z(S)$. Thus, since $R \subset S$, $R = S$ and we are done.

(VIII.18) THEOREM. (Fundamental Theorem of Galois Theory for L). Let $L = M_n(k)$, $k = GF(p^t)$ and $A = \mathrm{Aut}(L)$. Let

$$S^* = \{S \mid S \text{ is a simple subring of } L\}$$

$$G^* = \{G \mid G \text{ is a regular subgroup of } A\}.$$

Then

$$\mathrm{Aut}_A : S^* \to G^*$$

$$\mathrm{Rng}_L : G^* \to S^*$$

describes a Galois correspondence; i.e.,

(a) $\mathrm{Rng}_L \circ \mathrm{Aut}_A = \mathrm{id}_{S^*}$, $\mathrm{Aut}_A \circ \mathrm{Rng}_L = \mathrm{id}_{G^*}$

(b) Let G_1 and G_2 in G^* and S_1 and S_2 in S^* correspond under the above maps. Then

$$G_1 \leq G_2 \text{ if and only if } S_2 \subseteq S_1.$$

Proof. This follows directly from (VIII.16) and (VIII.17).

Section (D). The Units of $M_n(k)$.

Let k be a finite field and $M_n(k)$ be the n by n matrix ring over k . This section examines elementary results concerning the units of $M_n(k)$, i.e., the general linear group

$$GL_n(k) = \{A \text{ in } M_n(k) \mid A \text{ is invertible}\} .$$

The order of $GL_n(k)$ is obtained by counting the matrices in $M_n(k)$ having linearly independent rows.

(VIII.19) THEOREM. Let $|k| = q$. Then

$$|GL_n(k)| = (q^n - 1)(q^n - q) \cdots (q^n - q^{n-1})$$
$$= q^{n(n-1)/2} \prod_{i=1}^{n} (q^i - 1) .$$

The determinant of a matrix is a function $\det : M_n(k) \to k$ which when restricted to $GL_n(k)$ is a surjective group morphism $\det : GL_n(k) \to k^*$. Recall k^* is cyclic. Let $k^* = \langle \Theta \rangle$ and let d be a positive divisor of $|k^*| = q - 1$. Set

$$G_d = \{A \text{ in } GL_n(k) \mid \det(A) = \Theta^d\} .$$

The correspondence theorem for surjective group morphisms provides the following result.

(VIII.20) THEOREM.

(a) Let d be a positive divisor of $|k| - 1$. Then G_d is a normal subgroup of $GL_n(k)$ of order $|GL_n(k)|/d$.

(b) Let $|k| - 1 = p_1 \cdots p_t$ (p_i primes and not necessarily distinct).

Then the chain

$$GL_n(k) \supset G_{p_1} \supset G_{p_1 p_2} \supset \cdots \supset G_{p_1 \cdots p_{t-1}} \supset G_{p_1 \cdots p_t}$$

is the beginning of a composition chain in $GL_n(k)$.

Since $GL_n(k)$ is not a simple group, i.e., $GL_n(k)$ has proper normal subgroups, it is natural to ask, "What are the normal subgroups in $GL_n(k)$?" Above we have the beginning of an answer. In fact, the answer is reduced to a study of the normal subgroups of $G_{p_1 \cdots p_t}$. Let $G_{p_1 \cdots p_t}$ be denoted by $SL_n(k)$, i.e.,

$$SL_n(k) = \{A \text{ in } GL_n(k) \mid \det(A) = 1\} .$$

This group $SL_n(k)$ is called the <u>special linear group</u>.

To conclude the question of normal subgroups we examine the generators of $SL_n(k)$ and $GL_n(k)$.

For λ in k , I the identity matrix and E_{ij} the elementary matrix with 1 in the (i,j)-position and zeros elsewhere, let

$$B_{ij}(\lambda) = I + \lambda E_{ij} \qquad (i \neq j)$$

and

$$D(\lambda) = E_{11} + \cdots + E_{n-1,n-1} + \lambda E_{nn} .$$

The matrix $B_{ij}(\lambda)$ is called a <u>transvection</u> and $D(\lambda)$ is called a <u>dilation</u>.

<u>(VIII.21)</u> <u>LEMMA</u>. Let A be in $GL_n(k)$. Then A = (product of transvections) $\cdot$ $D(\lambda)$ where $\lambda = \det(A)$.

<u>Proof</u>. The action of multiplying A by $B_{ij}(\lambda)$ on the left is to effect a row operation, replacing the i-th row of A by the i-th row plus λ times the j-th row of A . Since A is invertible not every element in the first column of A is zero. Thus let $A = [a_{ij}]$ and assume $a_{i1} \neq 0$. Add $a_{i1}^{-1}(1 - a_{11})$ times the i-th row to the 1-st row. This places a 1 in the (1,1)-position. Now sweep out the first column and continue to the next column. Eventually, since these operations are performed by left multiplication by transvections, we have

$$B_1 \cdots B_s A = D(\lambda)$$

where B_i are transvections. Since $\det(B_{ij}(\alpha)) = 1$ and $\det(D(\lambda)) = \lambda$, we have $\det(A) = \lambda$. Finally, $B_{ij}(\alpha)^{-1} = B_{ij}(-\alpha)$ is a transvection. Hence

$$A = B_s^{-1} \cdots B_1^{-1} D(\lambda) .$$

<u>(VIII.22)</u> <u>THEOREM</u>.

(a) $GL_n(k)$ is generated by transvections and dilations.

(b) $SL_n(k)$ is generated by transvections.

Let $Z(G)$ denote the <u>center</u> of a group G, i.e.,

$$Z(G) = \{h \text{ in } G \mid hg = gh \text{ for all } g \text{ in } G\} .$$

(VIII.23) COROLLARY.

(a) $Z(GL_n(k)) = \{\alpha I \mid \alpha \text{ in } k^*\}$

(b) $Z(SL_n(k)) = \{\alpha I \mid \alpha \text{ in } k^*, \alpha^m = 1\}$.

Proof. If A is in the center of $GL_n(k)$ then A must commute with all transvections $B_{ij}(\lambda)$. This forces A to be a scalar multiple of the identity matrix.

To complete the question concerning normal subgroups of $SL_n(k)$ and in turn those of $GL_n(k)$, we now indicate that there are no proper normal subgroups of $SL_n(k)$ lying between $SL_n(k)$ and $Z(SL_n(k))$. Precisely, if N is a normal subgroup of $SL_n(k)$ not in the center then $N = SL_n(k)$. We only sketch the proof of the last statement. Details may be found in most texts on group theory.

The first step is to let $SL_n(k)$ act as a transformation group on itself by conjugation (inner-automorphism). One shows that the orbit or conjugacy class of any one transvection is the set of all transvections, i.e., $SL_n(k)$ is transitive on the set of transvections. Then, if N is a normal subgroup of $SL_n(k)$ properly containing the center, N

must contain a non-central element. Careful conjugation of this element produces a transvection in N. By the above comment we obtain all transvections in N and, by (VII.22), $SL_n(k) \subseteq N$. Hence $N = SL_n(k)$. To perform the matrix algebra one must have sufficient number of off-diagonal positions to work with. Thus, one finds the requirement that $n \geq 3$.

<u>(VIII.24)</u> <u>COROLLARY</u>. If $n \geq 3$ and N is a proper normal subgroup of $SL_n(k)$, then N is a subgroup of the center $Z(SL_n(k))$.

If $|k| = q$ where $q = p^t$ for p a prime, we now briefly examine the p-Sylow subgroups of $GL_n(k)$. This is the easiest case. A treatment of the remaining Sylow subgroups may be found in Dixon [11].

Let

$$T_n(k) = \{[a_{ij}] \mid a_{ii} \neq 0 \text{ and } a_{ij} = 0 \text{ for } i < j\} .$$

The subgroup $T_n(k)$ is called the <u>lower triangular group</u> and consists of all the A in $GL_n(k)$ which have all zeros above the diagonal. Let

$$ST_n(k) = \{[a_{ij}] \text{ in } T_n(k) \mid a_{ii} = 1 \text{ for } 1 \leq i \leq n\} .$$

By counting rows

$$|ST_n(k)| = q^{n-1} \cdot q^{n-2} \cdot \cdots \cdot q \cdot 1 = q^{n(n-1)/2} .$$

Thus $[GL_n(k):ST_n(k)] = \prod_{i=1}^{n} (q^i - 1)$ is not divisible by p. Hence $ST_n(k)$ is a p-Sylow subgroup of $GL_n(k)$.

It is easy to see that the normalizer of $ST_n(k)$ in $GL_n(k)$, i.e., $\{A \mid ATA^{-1}$ is in $ST_n(k)$ for all T in $ST_n(k)\}$, is $T_n(k)$. It is well-known the index of the normalizer of a p-Sylow subgroup in the whole group counts the p-Sylow subgroups. In this case,

$$[GL_n(k):T_n(k)] = (\prod_{i=1}^{n} q^i - 1)/(q - 1)^n$$

$$= \prod_{i=1}^{n} (1 + q + \cdots + q^{i-1}) .$$

<u>(VIII.25)</u> <u>THEOREM</u>. Let $|k| = p^t$ (p prime). Then

(a) $ST_n(k)$ is a p-Sylow subgroup of $GL_n(k)$.

(b) $T_n(k)$ is the normalizer of $ST_n(k)$.

(c) The number of p-Sylow subgroups in $GL_n(k)$ is

$$\prod_{i=1}^{n} (1 + q + \cdots + q^{i-1}) .$$

<u>Section (E)</u>. <u>Semi-Local Rings, The Standard Decomposition, and Algebras</u>.

This section concludes this chapter and deals first with rings R having $R/Rad(R)$ simple and second with finite dimensional algebras over a finite field.

Recall a ring R is <u>semi-local</u> if $R/\mathrm{Rad}(R)$ is a matrix ring $M_n(k)$ over a finite field k.

<u>(VIII.26)</u> <u>THEOREM</u>. (Structure Theorem for Semi-Local Rings). Let R be a semi-local ring. Then R is an n by n matrix ring over a local ring S. The integer n is unique and S is unique up to ring isomorphism.

<u>Proof</u>. Since R is semi-local, $R/\mathrm{Rad}(R) = M_n(k) = \overline{L}_1 \oplus \cdots \oplus \overline{L}_n$ where the $\overline{L}_i$ are minimal left ideals and $\overline{L}_i \simeq \overline{L}_j$. Apply (VII.12). Then ${}_R R = L_1 \oplus \cdots \oplus L_n$ (uniquely) where $L_i = Re_i$, $L_i \simeq L_j$ and the L_i are indecomposable. Apply (VII.3). This completes the proof.

For a ring R, let $1 = e_1 + \cdots + e_n$ where the e_i are mutually orthogonal idempotents. The <u>two-sided</u> <u>Pierce</u> <u>decomposition</u> of R relative to these idempotents is

$$R = e_1Re_1 \oplus \cdots \oplus e_nRe_n \oplus (\oplus \sum_{i \neq j} e_iRe_j)$$

(direct sum as groups).

<u>(VIII.27)</u> <u>THEOREM</u>. Let R be a ring. Then

$$R = S + N$$

where

(a) $S \cap N = 0$

(b) $S = \oplus\sum_{i=1}^{n} M_{n_i}(S_i)$ is a direct sum of n_i by n_i matrix rings over local rings S_i

(c) N is an additive subgroup of $\mathrm{Rad}(R)$

(d) $S/\mathrm{Rad}(S) \simeq R/\mathrm{Rad}(R)$.

Proof. Let $R/\mathrm{Rad}(R) = A_1 \oplus \cdots \oplus A_n$ where A_i is a matrix ring over a finite field k_i . Let $\bar{e}_i$ be the identity of A_i . Then the $\{\bar{e}_i\}$ may be lifted to a set of orthogonal idempotents $\{e_i\}$ in R with $1 = e_1 + \cdots + e_n$ and $e_i \to \bar{e}_i$ under $R \to R/\mathrm{Rad}(R)$. Set

$$S = e_1Re_1 \oplus \cdots \oplus e_nRe_n$$

and

$$N = \oplus \sum_{i \neq j} e_iRe_j .$$

Then $S \cap N = 0$, $R = S + N$ and N is a subgroup of $\mathrm{Rad}(R)$ since $N \equiv 0 \pmod{\mathrm{Rad}(R)}$.

The summand e_iRe_i is a ring with identity e_i in R and $e_iRe_i \to A_i$ under $R \to R/\mathrm{Rad}(R)$. By (VIII.26), e_iRe_i is an n_i by n_i matrix ring over a local ring S_i .

We call the decomposition of R described in (VIII.27) the <u>standard decomposition</u> of R .

We turn now to a brief examination of subalgebras of a matrix ring. For the remainder of this section fix the following setting:

> Let k be a finite field and A be a finitely generated (as a k-space) algebra over k . Assume $\dim_k(A) = m$ and $A/\mathrm{Rad}(A) = \oplus \sum_{i=1}^{n} M_{n_i}(k_i)$ (k_i finite fields).

Three basic results concerning A are provided in the following theorem.

<u>(VIII.28)</u> <u>THEOREM</u>. (For the above setting.)

(a) (Regular Representation). The algebra A is isomorphic to a subalgebra of $\mathrm{End}_k(A) \simeq M_m(k)$.

(b) (Skew-Polynomial Rings). The algebra A is a ring homomorphic image of a skew-polynomial ring

$$K[X_1,\ldots,X_t;\sigma_1,\ldots,\sigma_t]$$

where K is a finite extension field of k , the X_i are non-commuting indeterminates, the σ_i are k-automorphisms of K and $\alpha X_i = X_i\sigma_i(\alpha)$ for α in K and $1 \leq i \leq t$.

(c) (Wedderburn Principal Theorem). There exists a subalgebra B in A with

$$A = B + \mathrm{Rad}(A) ,$$

$$B \cap \mathrm{Rad}(A) = 0 ,$$

and $$B \simeq A/\mathrm{Rad}(A) .$$

Proof. To prove (a), let x be in A. Then x induces a k-linear mapping $A \to A$ by $a \to a^x$ where $a^x = ax$ (right multiplication) for all a in A. It is easy to check that $x \to (\ )^x$ is injective and additive. Further $a^{xy} = axy = (ax)^y = (a^x)^y$ and multiplication is preserved. Hence we have an injective algebra morphism $A \to \mathrm{End}_k(A)$.

To prove (b), let K be any field in A containing k and maximal with respect to $[K:k]$. Then A is a two-sided K-space. By (VIII.8) select a distinquished basis $\{b_1,\ldots,b_t\}$ For A over K. Define $K[X_1,\ldots,X_t;\sigma_1,\ldots,\sigma_t] \to A$ by $X_i \to b_i$.

The proof of (c) is slightly longer. We begin with a special case.

Let S be a local ring with characteristic p, maximal ideal m and residue field K. Let v be the nilpotency of m.

Let $K^* = \langle\bar{\theta}\rangle$ and let θ be a pre-image of θ in S. The commutative subring $T = (Z/Zp)[\theta]$ is a local subring of S with maximal ideal $N = m \cap T$. Observe $K = T/N$. Each element $\bar{a}$ of K has the form $\bar{a} = a + N$ for some a in T. Define $\phi : K \to T$ by $\phi : \bar{a} \to a^{p^v}$. It is straightforward to check that ϕ is a well-defined injective ring morphism. Thus T and hence S contains a copy of K as a subring.

Consequently, S is a two-sided K-space and if K is in the center then S is a K-algebra.

Since $S/m = K$ we have that $S = K + m$. Further each non-zero element of K in S is a unit. Hence $K \cap m = 0$.

Returning to the general case, let A have standard decomposition

$$A = (\oplus \sum_{i=1}^{n} M_{n_i}(S_i)) \oplus N$$

where the S_i are local with residue fields k_i. Since A has characteristic p, each S_i has characteristic p. By the above

$$S_i = k_i + m_i$$

where m_i is the maximal ideal of S_i. Then $M_{n_i}(S_i) = M_{n_i}(k_i) + M_{n_i}(m_i)$. Set $B = \oplus \sum_{i=1}^{n} M_{n_i}(k_i)$ in A and note

$\mathrm{Rad}(A) = \sum_{i=1}^{n} M_{n_i}(m_i) + N$.

<u>(VIII.1)</u> <u>EXERCISE</u>. (Smith). Show that if a ring R has a nilpotent non-zero element in its center then R is not semi-simple. Use this and show that if the characteristic of R is p^n $(n \geq 2)$, the R is not semi-simple. Give an example of a non-semi-simple ring which has no non-zero nilpotents in its center.

(VIII.2) EXERCISE. Show that if R is a commutative semi-simple ring then R is a direct sum of finite fields. Show that a commutative ring with no non-zero nilpotents is semi-simple.

(VIII.3) EXERCISE. If R is a simple ring show that every element of R is either a unit or a zero divisor.

(VIII.4) EXERCISE. (The matrix ring $M_n(k)$). Let k be a finite field and $M_n(k)$ the $n \times n$ matrix ring over k .

(a) Show $M_n(k)$ has only two ideals $M_n(k)$ and 0 .

(b) Show the center of $M_n(k)$ is naturally isomorphic to k .

(c) Let E_{ij} be the matrix having 1 in the (i,j)-position and zeros elsewhere. Show

(1) $\sum E_{ii} = I_n$, $E_{ij}E_{st} = \delta_{js}E_{it}$.

(2) $M_n(k)$ is a k-vector space of dimension n^2 with basis $\{E_{ij}\}$.

(3) If $J \subseteq \{1,\ldots,n\}$ and $E = \sum_{j \in J} E_{jj}$

(i) $EM_n(k)E$ is a ring

(ii) $M_n(k)E$ is a left ideal of $M_n(k)$ and a direct summand

(iii) $M_n(k) = M_n(k)E_{11} \oplus \cdots \oplus M_n(k)E_{nn}$ is a decomposition of $M_n(k)$ as minimal left ideals.

(d) Show that if $\Lambda : M_n(k) \to M_n(k)$ is a ring automorphism there exist an invertible matrix P and a field automorphism σ of k such that $\Lambda([a_{ij}]) = P^{-1}[\sigma(a_{ij})]P$ for all $A = [a_{ij}]$

in $M_n(k)$. (Hint: see the exercises in the last section.) Show that P is unique up to scalar multiplication. Show that $\wedge : M_n(k) \to M_n(k)$ is a k-algebra automorphism if and only if there is an invertible matrix P with $\wedge(A) = P^{-1}AP$ for all A in $M_n(k)$.

(VIII.5) EXERCISE. (Smith). If Λ is a k-algebra automorphism of $M_n(k)$ and $\Lambda(A) = P^{-1}AP$ for all A in $M_n(k)$ then the stablizer $I(\Lambda)$ of Λ is $\{A \mid \Lambda(A) = A\} = \{A \mid AP = PA\}$. Let $C(P) = \{A \mid AP = PA\}$, $m_P(X)$ be the minimal polynomial of P and $c_P(X)$ be the characteristic polynomial of P .

(a) Show $k[P] \cong k[X]/(m_P)$. Deduce that $k[P]$ is simple if and only if m_P is irreducible and $k[P]$ is semi-simple if and only if m_P is a product of distinct irreducibles.

(b) Show $C(P) = C(k[P])$. Show $C(P) = k[P]$ if and only if $c_P(X) = m_P(X)$.

(c) Show $C(k[P]) \cong \mathrm{Hom}_{k[P]}(V,V)$ (V = k-vector space of k-dimension n) .

(d) Show that if m_P is irreducible, then (by (a) $k[P]$ is a field)

(1) $[V:k] = [V:k[P]][k[P]:k]$ and $[V:k[P]] = n/\deg(m_P)$.

(2) $C(k[P])$ is a simple ring and $[C(k[P]):k] = [C(k[P]):k[P]][k[P]:k] = n^2/\deg(m_P)$.

(3) $C(P)$ is a field if and only if $m_P = c_P$.

(e) Suppose $m_P(X) = (X - c_1) \cdots (X - c_t)$ where the c_i's are distinct. Show $C(P)$ is a direct sum of simple k-algebras. If $m_P = c_P$ then $C(P)$ is a direct sum of $\deg(m_P)$ copies of k .

(f) (This involves a knowledge of canonical forms under similarity and decomposition of V under a linear map.) Show that if $m_P(X) = p_1(X)^{e_1} \cdots p_t(X)^{e_t}$ where the p_i are distinct irreducible monic polynomials and at least one $e_i > 1$, then $C(P)$ is not semi-simple.

(VIII.6) EXERCISE. (Carlitz-Hodges). (Continuation of the above.)

(a) Show $C(P) = k[P]$ if and only if $m_P = c_P$. We say that P is non-derogatory if $m_P = c_P$. Let $\deg(m_P) = n$.

(b) Let P be non-derogatory and $A = f(P)$ be in $C(P)$. Show A is invertible if and only if $(f(X), m_P(X)) = 1$. Show that to every such polynomial f there is a unique monic polynomial g with $\deg(g) = n$ and $(g, m_P) = 1$.

Thus the number of distinct invertible matrices A which commute with the non-derogatory matrix P is the number $\phi(m_P)$ of polynomials g of $\deg(g) < n$ such that $(g, m_P) = 1$. This is given by (II.19)

$$\phi(m_P) = |k|^m \prod_{f|m_P} (1 - \frac{1}{|k|^e})$$

where f runs through all monic irreducible divisors of m_P and $\deg(f) = e$. (Can you show this?)

(c) Show the number $Y(n)$ of $n \times n$ non-derogatory matrices in $M_n(k)$ is

$$Y(n) = |GL_n(k)| \sum_{\deg(f)=n} \frac{1}{\phi(f)}$$

where Σ extends over all monic polynomials f of degree n .

(VIII.7) EXERCISE. Let R be a ring. An additive map $d : R \to R$ is called a derivation if $d(rs) = rd(s) + d(r)s$. If $d : M_n(k) \to M_n(k)$ is a derivation show there is a P such that $d(A) = AP - PA$ for all A in $M_n(k)$.

(VIII.8) EXERCISE. (Walker). (Fermat's Theorem).

(a) Let A be a finitely generated algebra over the finite field k . Let $J(A)$ be the set of all f in $k[X]$ such that $f(a) = 0$ for all a in A . Show $J(A)$ is a principal ideal.

(b) Let $A = M_n(k)$. Show that $J(M_n(k)) = (f)$ where $f = f(n,p^m,X)$ $(|k| = p^m)$ is the monic least common multiple of all polynomials of degree n in $k[X]$.

(c) If B is a k-subalgebra of A , show

$$J(B) \supseteq J(A) .$$

(d) If $A_1,\ldots,A_t$ are k-algebras, show

$$J(\oplus\textstyle\sum_{i=1}^{t} A_i) = \bigcap_i J(A_i) .$$

(e) Show that

$$f(n,p^m,X) = (X^{p^{mn}} - X)f(n-1,p^m,X) .$$

Thus, show

$$f(m,p^n,X) = (X^{p^m} - X)(X^{p^{2m}} - X) \cdots (X^{p^{mn}} - X) .$$

(If $n = 1$ this is Fermat's Theorem.)

(VIII.9) EXERCISE. Regarded as a k-vector space, let $\sigma : M_n(k) \to M_n(k)$ be a k-linear map. If there exists a set $\{P_i, Q_i\}$ such that $\sigma(A) = \Sigma_i P_i A Q_i$ for all A in $M_n(k)$ then σ is called analytic. If $A_1, \ldots, A_r$ are k-free in $M_n(k)$ and $B_1, \ldots, B_r$ are arbitrary elements in $M_n(k)$, show there exists an analytic map σ with $\sigma(A_i) = B_i$, $1 \leq i \leq r$. (Hint: Use induction on r.)

(VIII.10) EXERCISE. (Brawley and Carlitz). Let R be a ring. Let $R[X_1, \ldots, X_m]$ be the polynomial ring in m non-commuting indeterminates. Further we do not assume the coefficients r in R commute with the X_i. Any such polynomial represents a function $R^m \to R$. Let $m \geq 2$ and show that every function from $R^m \to R$ is representable by a polynomial in $R[X_1, \ldots, X_m]$ if and only if $R = M_n(k)$ is a matrix ring over a finite field. (Compare with Theorem (II.4).)

(VIII.11) EXERCISE. Show $M_n(k) \otimes M_m(k) \simeq M_{mn}(k)$, k a finite field.

(VIII.12) EXERCISE. Let A and B be algebras over a finite field k.

(a) Let L be a proper ideal of A. Show $A \otimes B$ has a proper ideal $L \otimes B$.

(b) Show $A \otimes B$ is simple over k if and only if A and B are simple over k .

(c) If A is simple over k and $D = A \otimes B$, show $C_D(A) = B$.

(VIII.13) EXERCISE. Let A be an algebra over a finite field k . We have seen that right multiplication embeds A isomorphically in $End_k(A)$, i.e., $A \rightarrow A \subseteq End_k(A)$.

(a) Show that left multiplication by elements of A embeds the opposite algebra (i.e., the opposite ring) A^0 in $End_k(A)$.

(b) Show

(1) $C_{End(A)}(A) = A^0$

(2) $C_{End(A)}(A^0) = A$.

(VIII.14) EXERCISE. (Kruse and Price). Show the number of pairwise non-isomorphic semi-simple k-algebras of dimension m is less than $2^{m-1}e^{m/e}$ where $e = 2.71...$ is the base for the natural logarithms. (Hint: Consider partitions of m .)

(VIII.15) EXERICSE. (The General Linear Group). For a group G let $[G,G]$ denote the commutator of G . Show

(a) $[GL_n(k),GL_n(k)] = SL_n(k)$, $n \geq 3$

(b) $[SL_n(k),SL_n(k)] = SL_n(k)$, $n \geq 3$

(c) $[T_n(k),T_n(k)] = ST_n(k)$

(d) $T_n(k)$ is the normalizer of $ST_n(k)$ in $GL_n(k)$.

(Remark: For x and y in a group G, $[x,y] = x^{-1}y^{-1}xy$ is the <u>commutator</u> of x and y. Let $[G,G]$ denote the <u>commutator subgroup</u> of G generated by $\{x^{-1}y^{-1}xy \mid x,y \text{ in } G\}$. Hint: If i,j,t are distinct, then $[B_{ij}(\lambda),B_{jt}(\beta)] = B_{it}(\lambda\beta)$.)

<u>(VIII.16)</u> <u>EXERCISE</u>. (Recall Exercise (II.6).) Let k be a finite field. Show the group of k-automorphisms of the ring of fractions of $k[X]$,

$$X \to \frac{aX + b}{cX + d}$$

with $ad - bc = 1$ is isomorphic to $SL_2(k)/Z(SL_2(k))$ where $Z(SL_2(k))$ denotes the center of $SL_2(k)$.

<u>(VIII.17)</u> <u>EXERCISE</u>. Let Z denote the center of $SL_n(k)$ for k a finite field. Show

(a) $Z = \{\alpha I \mid \alpha^n = 1\}$

(b) $|Z| = d$ where $d = \text{g.c.d. } \{n, |k| - 1\}$.

<u>(VIII.18)</u> <u>EXERCISE</u>.

(a) Show that if $R = e_1Re_1 \oplus \cdots \oplus e_nRe_n \oplus N$ is a standard decomposition of R then $R = e_1Re_1 \oplus \cdots \oplus e_nRe_n$ if and only if the idempotents $\{e_i\}$ are commutative, i.e., e_i lie in the center of R.

(b) Show that if the standard decomposition of R is $R = eRe \oplus N$ then $R = eRe$.

(c) Suppose the radical of R is contained in the center of R.

Show R is isomorphic to a direct sum of local rings and matrix rings over finite fields.

(VIII.19) EXERCISE. (Nakayama). A left R-module M is called regular if there exist integers p and q such that $\oplus \sum_{i=1}^{p} M \simeq \oplus \sum_{j=1}^{q} R$. The rational number $t = q/p$ is called the rank of M .

(a) Let M be R-regular of rank t .

(i) Show that t is uniquely determined by M (use Krull-Schmidt Theorem).

(ii) If $S = \mathrm{End}_R(M)$, show that M is S-regular of rank t^{-1} .

(iii) If $S = \mathrm{End}_R(M)$, show $R = \mathrm{End}_S(M)$.

(b) Let S and R be rings with R a subring of S .

(i) If R is semi-local and S is R-regular, show that S is R-free.

(ii) (Artin-Whaples). If R is simple, show that S is R-free.

(VIII.20) EXERCISE. Let L be a left ideal in a ring R . Let b be in R such that $L + Rb = R$. Show that $L + b$ contains a unit.

(VIII.21) EXERCISE. If R is semi-local show that $M_n(R)$ is semi-local.

(VIII.22) EXERCISE. Let A be an algebra over a finite field k with $\mathrm{rad}(A) = N$. Show there is a decreasing sequence of algebras S_n, $n = 0,1,\ldots$ such that

$$S_n + N_{n-1} = S_{n-1}$$

$$S_n \cap N_{n-1} = N_n$$

where $N_n = N^{2^n}$.

(VIII.23) EXERCISE. Call an R-module M completely reducible if to every submodule K there is a complementary submodule L with $M = K \oplus L$. If $R = M_n(k)$ show every two-sided $M_n(k)$-module is completely reducible.

(VIII.24) EXERCISE. Let $\Lambda : M_n(k) \to M_n(k)$ be a ring automorphism of $M_n(k)$. Then $\Lambda(A) = PA^{\beta}P^{-1}$ for all A in $M_n(k)$. Identify $\Lambda \leftrightarrow \langle P,\beta \rangle$ with the pair $\langle P,\beta \rangle$ where P is in $GL_n(k)$ and β is in $\mathrm{Aut}(k)$. We know that β is unique and P is unique up to scalar multiple. Let $PGL_n(k) = GL_n(k)/(\text{Center of } GL_n(k))$. The group $PGL_n(k)$ is called the projective general linear group.

(a) Show there is a natural bijection between $\mathrm{Aut}(M_n(k))$ and $PGL_n(k) \times \mathrm{Aut}(k)$.

(b) Let $\Lambda \leftrightarrow \langle P,\beta \rangle$ and $\Gamma \leftrightarrow \langle Q,\theta \rangle$ be automorphisms. Show

$$\Lambda\Gamma \leftrightarrow \langle PQ^{\beta},\beta\theta \rangle .$$

(VIII.25) EXERCISE. (Continuation). Let H and K be groups. Suppose that for every h in H there exists an automorphism $\sigma_h : K \to K$ where $\sigma_h : t \to t^h = \sigma_h(t)$ satisfying

(1) $t^1 = t$

(2) $t^{h_1 h_2} = (t^{h_1})^{h_2}$.

(a) Show the set $G = K \times H$ forms a group, called the semi-direct product, under

$$\langle t_1, h_1\rangle\langle t_2, h_2\rangle = \langle t_1(t_2^{h_1}), h_1 h_2\rangle .$$

(b) Show $\mathrm{Aut}(M_n(k))$ is isomorphic to the semi-direct product of $PGL_n(k)$ and $\mathrm{Aut}(k)$.

(c) Show $PGL_n(k)$ is a normal subgroup of $\mathrm{Aut}(M_n(k))$ and there exists a natural exact sequence of groups

$$1 \to PGL_n(k) \to \mathrm{Aut}(M_n(k)) \to \mathrm{Aut}(k) \to 1 .$$

(VIII.26) EXERCISE. Let k be a finite field and V a k-space of dimension n . Let $L = \mathrm{End}_k(V)$ and S and T be simple k-algebras in L . If $\phi : S \to T$ is a ring isomorphism, show there exists a ring isomorphism $\Phi : L \to L$ with $\Phi|_S = \phi$; equivalently, there is a semi-linear map $\rho : V \to V$ with $\phi(s) = \rho s \rho^{-1}$ for all s in S . (This exercise and the next extend the Noether-Skolem Theorem (VIII.12). As a hint, parallel the proof of (VIII.12) and use the fact that when ϕ is restricted to the center of S subfields of S behave nicely.)

(VIII.27) EXERCISE. (Continuation of (VIII.26)). Let S be a simple ring in L. Let $\phi : S \to S$ be a ring isomorphism. Show there exists a semi-linear $\rho : V \to V$ with $\phi(s) = \rho s \rho^{-1}$ for every s in S. (Hint: Use the Galois Theory. Lift ϕ from S to Sk as follows. If $k_2 = k \cap Z(S)$ and $\phi|_{k_2} = \bar{\sigma}$, let σ be in $\mathrm{Aut}(k)$ such that $\sigma|_{k_2} = \bar{\sigma}$. Define $\bar{\phi} : Sk \to Sk$ by $\bar{\phi}(s\alpha) = \phi(s)\sigma(\alpha)$ for s in S and α in k and extend linearly. Then extend $\bar{\phi}$ by (VIII.26).)

(VIII.28) EXERCISE. Let S be a simple ring in $L = \mathrm{End}_k(V)$ where V is a k-space of finite dimension. Let $G = \mathrm{Aut}_A(S)$ where $A = \mathrm{Aut}(L)$. If G is normal in A, show $\mathrm{Aut}(S) \simeq \mathrm{Aut}(L)/G$. (Hint: For Φ in $\mathrm{Aut}(L)$, consider the group morphism $\Phi \to \Phi|_S$. Show $\Phi|_S : S \to S$, the kernel of the map is G, and that the map is onto $\mathrm{Aut}(S)$ by (VIII.26).)

(VIII.29) EXERCISE. Utilizing the above exercises, enumerate the isomorphism classes of simple subrings of $M_n(k)$ where k is a finite field. Determine the number of simple subrings in each isomorphism class.

IX. EXAMPLES: THE MATRIX RING AND THE GROUP RING

In this section we turn to the examples of the matrix ring and the group ring. First the radical of the matrix ring $M_n(R)$ is determined with regard to the radical of R. Second we examine the group ring in some detail. It is first shown that $R[G]$ is semi-simple if and only if both R is semi-simple and $|G|$ is a unit of R; and, then that $R[G]$ is local if R is a commutative local ring of characteristic p^t and G is a finite p-group.

Section (A). Matrix Rings.

If N is an ideal of a ring R then the matrix ring $M_n(R/N)$ is isomorphic to $M_n(R)/M_n(N)$. Further, every ideal of $M_n(R)$ is of the form $M_n(N)$ for an ideal N of R. Indeed, this gives an order preserving bijection between the lattice of ideals of $M_n(R)$ and the ideal lattice of R. This correspondence preserves ideal multiplication.

(IX.1) THEOREM. If R is a ring, then

$$\mathrm{Rad}(M_n(R)) = M_n(\mathrm{Rad}(R)) .$$

Proof. By the above P is a prime ideal of R if and only if $M_n(P)$ is a prime ideal of $M_n(R)$. Using (IV.8)

$$\mathrm{Rad}(M_n(R)) = \cap M_n(P) = M_n(\cap P) = M_n(\mathrm{Rad}(R)) .$$

(IX.2) COROLLARY. The matrix ring $M_n(R)$ is semi-simple if and only if R is semi-simple.

Section (B). The Group Ring.

Let G denote a finite multiplicative group and R a ring. Let $R[G]$ denote the set of formal sums

$$\sum_{g\varepsilon G} r_g g$$

where r_g is in R,

$$\Sigma r_g g + \Sigma s_g g = \Sigma (r_g + s_g)g ,$$

$$r(\Sigma r_g g) = \Sigma (rr_g)g ,$$

and

$$(\sum_{g\varepsilon G} r_g g)(\sum_{h\varepsilon G} s_h h) = \sum_{k\varepsilon G} t_k k$$

where $t_k = \sum_{gh=k} r_g s_h$.

With a subgroup H of G associate a left ideal $\omega(H)$ of $R[G]$ generated by all $1 - h$ for h in H. The mapping ω is injective from the lattice of subgroups of G to the lattice of left ideals of $R[G]$ and preserves inclusions. If H is a normal subgroup of G then $\omega(H)$ is a two-sided ideal of $R[G]$. Finally, if $\{h_i\}$ generates H then the left ideal $\omega(H)$ is generated by $\{1 - h_i\}$.

For a left ideal L of $R[G]$ let

$$\mathrm{Ann}(L) = \{r \text{ in } R \mid Lr = 0\}$$

denote the <u>right annihilator</u> of L.

<u>(IX.3)</u> <u>THEOREM</u>. The group ring $R[G]$ is semi-simple if and only if

(a) R is semi-simple

(b) $|G|$ is a unit in R.

<u>Proof</u>. Suppose $R[G]$ is semi-simple. Then (as rings) $R \approx R[G]/\omega(G)$. Thus R is semi-simple.

Let g be in G with order $0(g) = n$. Since $R[G]$ is semi-simple by (VIII.4) there is an r in $R[G]$ with $(1 - g)r(1 - g) = 1 - g$. Thus $(1 - g)(1 - r(1 - g)) = 0$ and $1 - r(1 - g)$ is in $\mathrm{Ann}(1 - g)$. By Exercise (IX.7) there is an s in $R[G]$ with

$$1 - r(1 - g) = (1 + g + \cdots + g^{n-1})s .$$

Under the <u>augmentation map</u> $\varepsilon : R[G] \to R$ by $\varepsilon : \Sigma\, r_g g \to \Sigma\, r_g$ (which is a ring morphism),

$$1 = n \cdot \varepsilon(s)$$

and n is a unit.

Let p be a prime dividing $|G|$. By Cauchy's Theorem there is an element g in G with $O(g) = p$. Thus p is a unit and hence $|G|$ is invertible in R.

Conversely, suppose $|G|$ is a unit and R is semi-simple. We need a lemma.

<u>(IX.4)</u> <u>LEMMA</u>. Let $|G| = n$ be a unit of R. Then a left R[G]-module is R[G]-projective if as an R-module it is R-projective.

<u>Proof</u>. Let P be a left R[G]-module. Since $R \subset R[G]$, P is an R-module. Suppose ${}_R P$ is R-projective and

$$\begin{array}{ccccc} & & {}_{R[G]}P & & \\ & & \big\downarrow \phi & & \\ {}_{R[G]}M & \xrightarrow{\ \Pi\ } & {}_{R[G]}N & \longrightarrow & 0 \end{array} \qquad \text{(row exact).}$$

There is an R-morphism $\psi : {}_R P \to {}_R M$. Let $|G| = n$ and p be in P.

Define

$$\phi' : {}_{R[G]}P \to {}_{R[G]}M$$

by

$$\phi'(p) = \frac{1}{n} \sum_{g \in G} g^{-1}\psi(gp) .$$

It is easy to show that if $h \in G$ then $\phi'(hp) = h\phi'(p)$ and that ϕ' is an $R[G]$-morphism. Further $\Pi\phi' = \phi$. Hence P is $R[G]$-projective.

To complete the proof of the theorem, note that ${}_R(R[G])$ is a finitely generated free R-module. Let a be in $R[G]$. We show that there is an a' with $aa'a = a$. Here it suffices to show $R[G]a$ is a direct summand of $R[G]$ as an $R[G]$-module. Since ${}_R(R[G]) \simeq \oplus \sum_{g \in G} R$ where R is semi-simple it is easy to show by induction that $(R[G])a \simeq \oplus \sum_{g \in G} H_g$ where H_g is a left ideal of R . But R is semi-simple and thus H_g is an R-direct summand of ${}_RR$. Hence $(R[G])a$ is an R-direct summand and hence a projective R-submodule of $R[G]$. By the lemma, $(R[G])a$ is projective as an $R[G]$-submodule of $R[G]$ and thus an $R[G]$-direct summand of $R[G]$. Hence $(R[G])a = (R[G])e$ where $e^2 = e$ and if $a'a = e$ then $aa'a = a$. Therefore by (VIII.4) $R[G]$ is semi-simple.

<u>(IX.5)</u> <u>COROLLARY</u>. (Maschke). If k is a finite field with $\chi(k) = p$ then the group algebra $k[G]$ is semi-simple if and only if $(p, |G|) = 1$.

Let S be a finite set. Suppose G is a finite p-group and that G

acts as a transformation group on S by $\sigma : s \to s^{\sigma}$ for s in S and σ in G. For s in S the <u>stablizer of s</u> is $I(s) = \{\sigma \text{ in } G \mid s^{\sigma} = s\}$ and the <u>orbit of s</u> is $[s] = \{s^{\sigma} \mid \sigma \text{ in } G\}$. Then $|[s]| = [G:I(s)]$ and since S is the disjoint union of its orbits we have

$$|S| = |S^G| + \Sigma_i p^{n_i} \qquad (n_i \geq 1)$$

where S^G denotes the fixed points of S under G. Hence

$$|S| \equiv |S^G| \pmod{p} .$$

As an aside, $|G| = p^n$, thus if G acts on itself as a transformation group by conjugation, then G^G is the center of G. Since the center of G is non-empty we see that it must contain at least p elements.

<u>(IX.6)</u> <u>THEOREM</u>. Let R be a commutative local ring of characteristic p^t. Let G be a finite p-group. Then $R[G]$ is a local ring.[1]

<u>Proof</u>. Let m denote the maximal ideal of R and $k = R/m$. Then, by (V.3) $m(R[G]) \subseteq \mathrm{Rad}(R[G])$. Thus it suffices to show $R[G]/m(R[G]) \cong k[G]$ is a local ring. Let $I = \ker(\varepsilon)$ where ε is the augmentation

---

[1]See Exercise (IX.12).

morphism (see proof of (IX.3)). Then $k \simeq k[G]/I$. Since k is a simple module, I is a maximal left ideal of $k[G]$. Further if k is the only simple module $I = \mathrm{Rad}(k[G])$ and $k[G]$ is local.

Let M be a simple non-zero $k[G]$-module. Then G acts as a transformation group on M and

$$|M^G| \equiv |M| \pmod p .$$

Since $|M|$ is a power of p (since it is a k-vector space) we have $|M^G| \neq 0$. Thus M^G is a non-zero $k[G]$-submodule of M. Since M is simple $M^G = M \simeq k[G]/I \simeq k$.

<u>(IX.7)</u> <u>COROLLARY</u>. (Dickson). Let k be a finite field of characteristic p. Let G be a p-group. Then $k[G]$ is a local ring with $\mathrm{Rad}(k[G]) = \omega(G)$.

<u>(IX.1)</u> <u>EXERCISE</u>. (Morita Duality for $M_n(R)$). Let R be a ring, let S denote $M_n(R)$, $\{E_{ij}\}$ denote the standard matrix units of S, and ${}_R\mathrm{Mod}$ and ${}_S\mathrm{Mod}$ denote the collections of finitely generated R and S-modules, respectively. Define maps $F: {}_R\mathrm{Mod} \to {}_S\mathrm{Mod}$ and $G: {}_S\mathrm{Mod} \to {}_R\mathrm{Mod}$ as follows:

(a) For M in ${}_S\mathrm{Mod}$ let $G(M) = E_{11}M$. If r is in R, let rI denote the scalar matrix. The action of R on $E_{11}M$ is

given by

$$r(E_{11}M) = (rI)(E_{11}M) .$$

(b) For M in ${}_R\text{Mod}$ let $F(M) = \oplus \Sigma_{i=1}^{n} M$. If $A = [r_{ij}]$ is in S , the action of A on $F(M)$ is defined by

$$[r_{ij}](m_1,\ldots,m_n) = (\Sigma_j r_{1j}m_j,\ldots,\Sigma_j r_{nj}m_j) .$$

We map morphisms as follows:

(a) If $\phi : M \to N$ is an S-morphism in ${}_S\text{Mod}$, let

$$G(\phi) : G(M) \to G(N)$$

be given by

$$G(\phi)(E_{11}m) = E_{11}\phi(m) .$$

(b) If $\phi : M \to N$ is an R-morphism in ${}_R\text{Mod}$, let

$$F(\phi) : F(M) \to F(N)$$

be given by

$$F(\phi)(m_1,\ldots,m_n) = (\phi(m_1),\ldots,\phi(m_n)) .$$

Then,

(1) If M is in ${}_S\text{Mod}$, show $G(M)$ is in ${}_R\text{Mod}$.
If M is in ${}_R\text{Mod}$, show $F(M)$ is in ${}_S\text{Mod}$.

(2) For morphisms in ${}_R\text{Mod}$, show

(i) $F(id_M) = id_{F(M)}$ (id = identity)

(ii) $F(\alpha\beta) = F(\alpha)F(\beta)$

(iii) $F(\alpha + \beta) = F(\alpha) + F(\beta)$.

Show analogous statements are true for G . (This shows that F and G are "functors.")

(3) Show that for every M in ${}_R\text{Mod}$ and N in ${}_S\text{Mod}$

$$M \simeq GF(M) \qquad \text{(as R-modules)}$$
$$N \simeq FG(N) \qquad \text{(as S-modules)}$$

where GF and FG are the obvious compositions.

(4) Show that for every M and N in ${}_R\text{Mod}$,

$$F : \text{Hom}_R(M,N) \to \text{Hom}_S(F(M),F(N))$$

by

$$F : \phi \to F(\phi)$$

is an additive group isomorphism and if $M = N$ it is a ring isomorphism. Show the analogous statement is true for G .

(5) Let $M \xrightarrow{\alpha} N \xrightarrow{\beta} P$ be an exact sequence of modules and morphi[sms] in ${}_R\text{Mod}$. Show that $F(M) \xrightarrow{F(\alpha)} F(N) \xrightarrow{F(\beta)} F(P)$ is exact in ${}_S\text{Mod}$. Show the analogous statement is true for G .

(6) Show that M is projective in ${}_R\text{Mod}$ (${}_S\text{Mod}$) if and only if $F(M)$ $(G(M))$ is projective in ${}_S\text{Mod}$ (${}_R\text{Mod}$).

(7) Show that as rings

$$R^0 \simeq \text{End}_R({}_RR) \simeq \text{End}_S(\oplus\Sigma_{j=1}^{n} R)$$
$$S^0 \simeq \text{End}_S({}_SS) \simeq \text{End}_R(\oplus\Sigma_{j=1}^{n} R) .$$

(8) Show that there is a lattice preserving bijection between th[e] ideals of R and those of $M_n(R)$ given by

$$A \to M_n(A) \ .$$

Show this bijection preserves ideal multiplication.

(IX.2) EXERCISE. Let R be a commutative ring, A an element in $M_n(R)$.

(a) Show the following are equivalent.

(1) A is invertible.

(2) $\det(A)$ is a unit of R.

(b) Show the following are equivalent.

(1) A is a zero divisor.

(2) $\det(A)$ is a zero divisor of R.

(IX.3) EXERCISE. Let $R = M_n(S)$ and $\sigma : R \to T$ be a surjective ring morphism. Show that T is $M_n(U)$ and there exists a ring morphism $\sigma' : S \to U$ which induces σ. (Hint: consider the matrix units.)

(IX.4) EXERCISE. Suppose R is a ring which has all left ideals principal. Show that all left ideals of $M_n(R)$ are principal.

(IX.5) EXERCISE. Let R be a ring. Suppose that R is generated by the subset S (as a ring). Show that $M_n(R)$ is generated by

$$Y = \begin{bmatrix} 0 & 1 & 0 & \cdots & 0 \\ 0 & 0 & 1 & \cdots & 0 \\ \cdots & & & & \\ 0 & 0 & 0 & \cdots & 1 \\ 1 & 0 & 0 & \cdots & 0 \end{bmatrix}$$

and the set $\{sE_{11} \mid s \text{ in } S\}$. (Hint: see (VIII.5).)

(IX.6) EXERCISE. Let R be a ring and M an R-module generated by n elements. Show that M corresponds to a cyclic $M_n(R)$-module under the Morita duality described in the earlier exercise. (This reduces some questions on finitely generated modules to cyclic modules.)

(IX.7) EXERCISE. (See notation before Theorem (IX.3)). Show

(a) For a subgroup H of G

$$\mathrm{Ann}(\omega(H)) = (\sum_{h \in H} h)R .$$

(b) For g in G , if the order of g , 0(g) , is n then

$$\mathrm{Ann}(1 - g) = (1 + g + \cdots + g^{n-1})R .$$

(IX.8) EXERCISE. Show the following:

(a) Let S be a subring of R . Then

$$S[G] \cap \mathrm{Rad}(R[G]) \subset \mathrm{Rad}(S[G]) .$$

(We have equality if S is in the center of R .)

(b) Let H be a subgroup of G . Then

$$R[H] \cap \mathrm{Rad}(R[G]) \subset \mathrm{Rad}(R[H]) .$$

(We have equality if H is in the center of G .)

(c) $R \cap \mathrm{Rad}(R[G]) = \mathrm{Rad}(R)$.

(IX.9) EXERCISE. Let G and H be finite Abelian p-groups and k a finite field of characteristic p . If $k[G] \simeq k[H]$ show $G \simeq H$. (Hint: Use induction on $|G| = |H|$ and note that $(k[G])^p = k^p[G^p]$.)

(IX.10) EXERCISE. Let G be a finite group and R be a ring. If $G = G_1 \times G_2$, does $R[G] = R[G_1] \oplus R[G_2]$? If $R = R_1 \oplus R_2$, does $R[G] = R_1[G] \oplus R_2[G]$?

(IX.11) EXERCISE. (Coleman). Let G be a group of order n and k a finite field of characteristic p such that $(p,n) = 1$. Let N be a normal subgroup of G and $\omega(N)$ denote the ideal generated by the set of all $1 - h$ where h is in N. Show $k[G] \approx k[G/N] \oplus \omega(N)$.

(IX.12) EXERCISE. (This improves (IX.6)). Let R be a local ring and G be a group of order $n > 1$. Show the following are equivalent.

(a) $R[G]$ is local.

(b) R has characteristic p^t and $n = p^s$.

When the above occurs, show that $\mathrm{Rad}(R[G])$ is $\sum_{\substack{\sigma \in G \\ \sigma \neq 1}} (1 - \sigma)R + m[G]$

where m is the maximal ideal of R.

X. BASIC RINGS AND CHECKERED MATRIX RINGS

The section (VIII), this section, and the next section provide the basic structure theory of arbitrary finite rings. Here, as in (VIII), we lift idempotents and show that a finite ring may be realized as a checkered matrix ring over a basic ring. Using this approach generally a problem concerning a finite ring may be first viewed as a question about basic finite rings and second as a question in matrix theory over these basic rings.

A ring S is called basic or reduced if $S/\mathrm{Rad}(S)$ is a direct sum of finite fields. This section will illustrate how the theory of finite rings may be reduced to the theory of finite basic rings and the theory of certain matrix rings. If S is basic then the number of finite fields appearing in the decomposition of $S/\mathrm{Rad}(S)$ is called the height of S .

Let S be a basic ring of height n . Then S has a set of n mutually orthogonal non-zero idempotents summing to 1 and in any set of mutually orthogonal non-zero idempotents there are no more than n idempotents. Suppose $\{e\} = \{e_1,\ldots,e_n\}$ is a system of orthogonal idempotents in S and suppose $\{q\} = \{q_1,\ldots,q_n\}$ is a set of n positive integers. The

<u>checkered matrix ring</u> $M(S,\{e\},\{q\})$ is the ring of all matrices of the form

$$\begin{bmatrix} A_{11} & A_{12} & \cdots & A_{1n} \\ A_{21} & A_{22} & \cdots & A_{2n} \\ \vdots & \vdots & \ddots & \vdots \\ A_{n1} & A_{n2} & \cdots & A_{nn} \end{bmatrix}$$

where A_{ij} is a q_i by q_j block matrix with elements in e_iSe_j .

<u>(X.1)</u> <u>THEOREM</u>. Let R be a ring. Then there exists a basic ring S , a set of orthogonal idempotents $\{e\} = \{e_1,\ldots,e_n\}$ and a set of positive integers $\{q\} = \{q_1,\ldots,q_n\}$ such that R is isomorphic (as a ring) to the checkered matrix ring $M(S,\{e\},\{q\})$.

<u>Proof</u>. We have a decomposition of R as left ideals

$${}_RR = \oplus\Sigma_{i=1}^{n}\Sigma_{j=1}^{q_i} L_i^{(j)}$$

where the $L_i^{(j)}$ are left ideals, $L_i^{(j)} \simeq L_k^{(t)}$ if and only if $i = k$, and $L_i^{(j)} = Re_i^{(j)}$ with

$$1 = \Sigma_{i=1}^{n}\Sigma_{j=1}^{q_i} e_i^{(j)} .$$

Denote by H_i the sum $\oplus\Sigma_{j=1}^{q_i} L_i^{(j)}$.

Then (as rings)

$$R^0 \simeq \mathrm{End}_R({}_R R)$$

$$= \mathrm{End}_R(\oplus \Sigma_i \; H_i)$$

which (by VII.1) is isomorphic to the ring of matrices

$$\left\{ \begin{bmatrix} \phi_{11} & \cdots & \phi_{1n} \\ \vdots & \ddots & \vdots \\ \phi_{n1} & \cdots & \phi_{nn} \end{bmatrix} \;\middle|\; \phi_{ki} \text{ in } \mathrm{Hom}_R(H_i, H_k) \right\} .$$

Now, setting $L_i = L_i^{(1)}$, we have $L_i \simeq L_i^{(j)}$, $1 \leq j \leq q_i$. Thus (as R-modules)

$$H_i \simeq \oplus \Sigma L_i \qquad (q_i \text{ summands}) .$$

Thus (as Abelian groups)

$$\mathrm{Hom}_R(H_i, H_k) \simeq \mathrm{Hom}_R(\oplus \; \Sigma L_i , \oplus \Sigma \; L_k)$$

$$= \mathrm{Hom}_R(\oplus \; \Sigma Re_i , \oplus \Sigma \; Re_k)$$

where $e_i = e_i^{(1)}$ and $e_k = e_k^{(1)}$. Then, employing an argument analogous to the proof of (VII.1), $\mathrm{Hom}_R(H_i, H_k)$ is isomorphic as a group to the set of q_i by q_k matrices over $\mathrm{Hom}_R(Re_i, Re_k) \simeq e_i Re_k$.

Now set $S = eRe$ where $e = e_1 + \cdots + e_n$. Then S is a subring of R with identity e . Note that $e_i = ee_i = e_i e$, $1 \leq i \leq n$, and thus $e_i Re_j = e_i eRee_j = e_i Se_j$.

We next claim that S is a basic ring. If s is in S ,

$$s = ese = \sum_{i,j=1}^{n} e_i s e_j .$$

Observe that $e_i S e_j \equiv 0 \bmod \text{Rad}(S)$ when $i \neq j$ ($\text{Rad}(S) = S \cap \text{Rad}(R)$ by (VII.4)). Hence

$$S \equiv \Sigma_{i=1}^{n} e_i S e_i \quad \bmod \text{Rad}(S)$$
$$\equiv \Sigma_{i=1}^{n} e_i R e_i \quad \bmod \text{Rad}(R) .$$

Since e_i is minimal, $e_i R e_i$ is a finite field modulo Rad(R) and S is basic.

We have shown that R^0 is isomorphic to a checkered matrix ring $M(S,\{e\},\{q\})$. We obtain a checkered matrix ring isomorphic to R by taking the transpose of each element of $M(S,\{e\},\{q\})$.

REMARKS AND NOTATION. The anti-isomorphism $R \to M(S,\{e\},\{q\})$ is denoted by ρ . We employ the notation introduced in the proof of (X.1) through the remainder of this section. Thus, $S = eRe$ is a subring of R with identity e where $e = \Sigma_{i=1}^{n} e_i^{(1)}$ and n is the height of S . The ring S is called the basic ring associated with R . The uniqueness theorems of (VII) and (VIII) imply that if there also exists an anti-isomorphism $R \to M(\overline{S},\{\overline{e}\},\{\overline{q}\})$ where $\overline{S}$ is basic of height m then $m = n$ and after relabeling there is an isomorphism $\psi : \overline{S} \to S$ with $\psi\overline{e}_i = e_i$ and $\overline{q}_i = q_i$ for $1 \leq i \leq n$. A careful examination of $\rho : R \to M(S,\{e\},\{q\})$ shows that if H is an ideal of R then H is carried by ρ to the set of matrices $\{[A_{ij}]\}$ where A_{ij} is a q_i by q_j matrix with elements r_{ij} in $e_iSe_j \cap H = e_iRe_j \cap H$.

(X.2) THEOREM. Let R be a ring with associated basic ring S . Then there exists a natural bijection between the ideals of R and those of S . The bijection is a lattice isomorphism and preserves ideal multiplication.

Proof. Let H be an ideal of R . As above we identify S with a subring of R . Define the correspondence by

$$H \to L$$

where $L = H \cap S = eHe$ and $H = RLR$.

The anti-isomorphism $\rho : R \to M(S,\{e\},\{q\})$ maps H onto the set of matrices composed of blocks $A_{ij} = [r_{ij}]$ with r_{ij} in $e_iSe_j \cap H = e_iRe_j \cap H$. But $e_iHe_j = H \cap e_iSe_j = L \cap e_iSe_j = e_iLe_j$. This shows that H is uniquely determined by L and $H \to L$ is injective.

Let L be an ideal of S . Then $H = RLR$ is an ideal of R . Further, since $Le = L = eL$, $eHe = eRLRe = eReLeRe = SLS = L$. Thus the correspondence is surjective.

The map is clearly a lattice isomorphism.

Finally, suppose L_1 and L_2 are ideals in S. Then $L_1e = L_1$ and $eL_2 = L_2$. Thus $L_1RL_2 = L_1eReL_2 = L_1SL_2 = L_1L_2$. That is, $H_1H_2 = (RL_1R)(RL_2R) = RL_1L_2R$.

Continuing with the notation in the proof of (X.1), the rings

$$T_1 = e_1Re_1, \ldots, T_n = e_nRe_n$$

are the <u>local rings associated with</u> R and are unique up to ring isomorphism. But $T_i = e_iRe_i = e_iSe_i$. Thus

<u>(X.3)</u> <u>THEOREM</u>. A ring R and its associated basic ring S have the same associated local rings.

<u>(X.1)</u> <u>EXERCISE</u>. Let R be a ring and ${}_RR = \oplus \Sigma_{i=1}^{n} L_i$ where the L_i are indecomposable left ideals. Show that R is a basic ring if and only if whenever $L_i \simeq L_j$ (as R-modules) then $i = j$.

<u>(X.2)</u> <u>EXERCISE</u>. (Brauer). Let R be a ring with basic ring S. Suppose S has $\{e\} = \{e_1, \ldots, e_n\}$ as a complete set of orthogonal non-zero idempotents, ψ is a ring automorphism of S leaving all the e_i fixed, and $R = M(S, \{e\}, \{q\})$.

(a) Show that ψ induces a natural ring automorphism $\overline{\psi}$ on R.

(b) Let Θ be a ring automorphism of R. Show that there is a ring automorphism ψ of S and a unit u of R such that

$$\Theta(r) = u(r)^{\overline{\psi}}u^{-1}$$

for every r in R ; i.e., Θ is the composition of $\overline{\psi}$ and an inner-automorphism of R .

(Hint: See Exercise (VII.11).)

(X.3) EXERCISE. Let n be the number of idempotents in a ring R . Sho

(a) n is odd (we are considering only non-zero idempotents)

(b) If $n = 3$, then R is a direct sum of two local rings.

(X.4) EXERCISE. Let $R = M(S,\{e\},\{q\})$ be a checkered matrix ring with basic ring S . Determine the center of R .

(X.5) EXERCISE. (Osima). Let M be an R-module, T be $\mathrm{End}_R(M)$ and S be the basic ring of T (it is assumed that M is finitely generated) Suppose that

$$(*) \qquad M = \oplus \sum_{i=1}^{n} \sum_{j=1}^{q_i} M_i^{(j)}$$

is a direct sum decomposition of M into indecomposable R-modules $M_i^{(j)}$ where $M_i^{(j)} \simeq M_t^{(k)}$ (as R-modules) if and only if $i = t$.

(a) Show that an idempotent e of T is minimal if and only if Me is an indecomposable left R-module.

(b) Show that if e and f are idempotents of T then $e \simeq f$ (see exercises in (VII)) if and only if $Me \simeq Mf$ (as R-modules)

(c) Show $\text{End}_R(Me) = e(\text{End}_R(M))e$ for an idempotent e of T.

(d) Show that corresponding to (*) there is a decomposition of 1 in T as

$$1 = \Sigma_i \ \Sigma_j \ e_i^{(j)}$$

where the $e_i^{(j)}$ are orthogonal minimal idempotents. Let $e = \Sigma_{i=1}^n \ e_i^{(1)}$.

(1) Show that $S = eTe$.

(2) Show $\text{End}_R(Me) = S$.

(X.6) EXERCISE. (Morita). (Morita Duality for Basic Rings). (See exercise on Morita Duality for $M_n(R)$ in (IX)). Let R be a ring with

$$1 = \Sigma_{i=1}^n \ \Sigma_{j=1}^{q_i} \ e_i^{(j)}$$

a decomposition of 1 into orthogonal minimal idempotents such that $Re_i^{(j)} \simeq Re_t^{(k)}$ if and only if $i = t$. Set $e = \Sigma_{i=1}^n \ e_i^{(1)}$ and $S = eRe$ (basic ring of R).

(a) Note that Re is a two-sided (R,S)-module. Show

(1) Re is a projective left R-module.

(2) Re is a projective right S-module.

(3) Any S-morphism of Re is given by left multiplication by an element of R.

(4) Any R-morphism of Re is given by right multiplication by an element of S.

(b) (1) Let X be a left R-module. Define a mapping

$$T_1 : {}_R\text{Mod} \to {}_S\text{Mod}$$

by

$$T_1(X) = eR \otimes_R X \qquad (\otimes_R \text{ denotes the tensor product}^1 \text{ over } R).$$

If X and Y are R-modules and $\sigma : X \to Y$ is an R-morphism define

$$T_1(\sigma) : T_1(X) \to T_1(Y)$$

by

$$T_1(\sigma)\ (t \otimes x) = t \otimes \sigma(x) .$$

Show that

(i) If X is in ${}_R\text{Mod}$, $T_1(X)$ is in ${}_S\text{Mod}$.

(ii) $T_1(id_X) = id_{T_1(X)}$, $T_1(\sigma + \beta) = T_1(\sigma) + T_1(\beta)$ and $T_1(\sigma\beta) = T_1(\sigma)T_1(\beta)$ for R-morphisms σ, β and the identity morphism id_X.

(Recall: ${}_R\text{Mod}$ (${}_S\text{Mod}$) denotes the collection of finitely generated left R-modules (left S-modules).)

(2) Let Y be a left S-module. Define a mapping

$$T_2 : {}_S\text{Mod} \to {}_R\text{Mod}$$

by

---

[1] The "tensor product" for algebras over a field is given in (VIII). For the general case see the exercises in (XII).

$$T_2(Y) = \mathrm{Hom}_S(eR,Y) .$$

If X and Y are S-modules and $\sigma : X \to Y$ is an S-morphism define

$$T_2(\sigma) : T_2(X) \to T_2(Y)$$

by

$$T_2(\sigma)(f) = \sigma f .$$

Show that

(i) For Y in ${}_S\mathrm{Mod}$, $T_2(Y)$ is in ${}_R\mathrm{Mod}$.

(ii) $T_2(id_Y) = id_{T_2(Y)}$, $T_2(\sigma + \beta) = T_2(\sigma) + T_2(\beta)$ and $T_2(\sigma\beta) = T_2(\sigma)T_2(\beta)$ for S-morphisms σ,β and the identity morphism id_Y.

(c) Show that for every M in ${}_R\mathrm{Mod}$ and every N in ${}_S\mathrm{Mod}$

$$M \simeq T_2T_1(M) \qquad \text{(as R-modules)}$$
$$N \simeq T_1T_2(N) \qquad \text{(as S-modules)} .$$

(Hint: Use (a) and show

$$\begin{aligned} M \simeq R \otimes_R M \simeq \mathrm{Hom}_S(eR,eR) \otimes_R M \\ \simeq \mathrm{Hom}_S(eR,eR \otimes_R M) \\ = T_2T_1(M) \end{aligned}$$

where the isomorphism

$$\tau : \mathrm{Hom}_S(eR,eR) \otimes M \to \mathrm{Hom}_S(eR,eR \otimes_R M)$$

is given by

$$[\tau(\alpha \otimes m)](x) = \alpha(x) \otimes m \ .$$

For the other isomorphism, show that

$$N \simeq \mathrm{Hom}_S(S,N) \simeq \mathrm{Hom}_S(\mathrm{Hom}_R(eR,eR),N) \simeq eR \otimes_R \mathrm{Hom}_S(eR,N) = T_1T_2(N)$$

where

$$\sigma_N : eR \otimes_R \mathrm{Hom}_S(eR,N) \to \mathrm{Hom}_S(\mathrm{Hom}_R(eR,eR),N)$$

is given by

$$[\sigma_N(r \otimes \alpha)](\beta) = \alpha(\beta(r)) \ .)$$

(d) For M and N in ${}_R\mathrm{Mod}$ show that

$$T_1 : \mathrm{Hom}_R(M,N) \to \mathrm{Hom}_S(T_1(M),T_1(N))$$

is an additive group isomorphism and if $M = N$ it is a ring isomorphism. Show the analogous statement is true for

$$T_2 : \mathrm{Hom}_S(M,N) \to \mathrm{Hom}_R(T_2(M),T_2(N)) \ .$$

(e) Show that T_1 (respectively, T_2) preserves exact sequences and projective modules.

<u>REMARK</u>. Morita [36]. The above results may be extended to show that if the basic ring S of R has any of the following properties then R has also the same property; (1) quasi-Frobenius, (2) generalized uni-serial, (3) uni-serial, (4) primary decomposable, (5) primary (semi-local), (6) weakly symmetric, (7) almost symmetric, (8) semi-simple, (9) simple, (10) cleft. Thus the structure of R is intimately related to the structure of its basic ring S .

(X.7) EXERCISE. (Hannula [31]). Let R be a ring with basic ring S.

(a) Show there exists an integer m and an idempotent e in the ring $M_m(S)$ of m by m matrices over S such that $R = eM_m(S)e$ and $M_m(S) = M_m(S)eM_m(S)$.

(b) Let e be on idempotent of R with $ReR = R$. Show a basic subring of eRe is isomorphic to S.

(c) Let R have idempotents e and f and suppose that eRe and fRf both contain the basic subring S. If $\sigma : eRe \to fRf$ is a ring isomorphism which fixes S, show σ is induced by an inner automorphism of R.

[Remark: For the application of the above exercise in the construction of quasi-Frobenius (self-injective) rings see Hannula [31]].

XI. DECOMPOSITION OF A RING AS IDEALS

The last section examined an arbitrary finite ring R from the standpoint of the lifting of idempotents from $R/\mathrm{Rad}(R)$ and, consequently, the decomposition of R as a direct sum of left ideals. This section concerns the decomposition of R as a direct sum of two-sided indecomposable ideals. This is nicely visualized from the standpoint of basic rings and checkered matrix rings and is approached in this fashion.

Let L be a left ideal of a ring R. A Jordan-Hölder chain for L is a chain of left ideals

$$L = L_1 \supset L_2 \supset \cdots \supset L_n \supset L_{n+1} = 0$$

such that

$$L_i/L_{i+1}$$

is a simple non-zero R-module for $1 \leq i \leq n$. The quotient module L_i/L_{i+1} of the Jordan-Hölder chain is called a chain factor and the integer n is called the length of the chain. It is well-known [see (III)] that such chains exist and if

$$L = L_1 \supset L_2 \supset \cdots \supset L_n \supset L_{n+1} = 0$$

$$L = L'_1 \supset L'_2 \supset \cdots \supset L'_m \supset L'_{m+1} = 0$$

are two Jordan-Hölder chains for L then $m = n$ (the lengths are the same) and there is a permutation σ of $\{1,\ldots,n\}$ such that $L_i/L_{i+1} \simeq L'_{\sigma(i)}/L'_{\sigma(i)+1}$ (as R-modules) for $1 \leq i \leq n$.

The last section emphasized the decomposition of a ring via its left ideal structure. In this section we explore the decomposition of R as a direct sum of ideals.

<u>(XI.1)</u> <u>THEOREM</u>. Let L be a left ideal in R and e be a minimal idempotent in R . Then L has a chain factor isomorphic to $Re/(Rad(R))e$ (as R-modules) if and only if $eL \neq 0$.

<u>Proof</u>. Let

$$L = L_1 \supset L_2 \supset \cdots \supset L_n \supset L_{n+1} = 0$$

be a Jordan-Hölder chain for L . Suppose $eL \neq 0$. First observe that if $eL_i \subset L_{i+1}$ for all i then $eL = e^nL \subset L_{n+1} = 0$. Thus there is an i with $eL_i \not\subset L_{i+1}$ and consequently a chain factor $M = L_i/L_{i+1}$ with $eM \neq 0$. Thus for some m in M , $em \neq 0$ and $Rem \neq 0$. Since M is simple, $Rem = M$. We hence have a natural surjective R-morphism $Re \to M$ by $re \to rem$. Thus $M \simeq Re/N$ where N is a maximal left R-submodule of Re . But Re (see comment before (VII.10)) has only one maximal submodule $N = (Rad(R))e$. Thus $Re/(Rad(R))e \simeq L_i/L_{i+1}$.

Conversely, suppose we have an R-module isomorphism $Re/(Rad(R))e \to L_i/L_{i+1}$.

Then the image of $x = e + (\mathrm{Rad}(R))e$ is $r + L_{i+1}$ where r is in L_i and not in L_{i+1} . But $ex = x$ thus er is not in L_{i+1} and hence $er \neq 0$. Therefore $eL \neq 0$.

<u>(XI.2)</u> <u>COROLLARY</u>. Let f be an idempotent and e be a minimal idempotent. Then Rf has a chain factor R-isomorphic to $Re/(\mathrm{Rad}(R))e$ if and only if $eRf \neq 0$.

Let H be an ideal of R . Then H is said to be <u>indecomposable</u> if whenever $H = A \oplus B$ where A and B are ideals then $A = 0$ or $B = 0$. Since R is finite it is clear that

$$R = H_1 \oplus \cdots \oplus H_n$$

where the H_i are indecomposable ideals. It is our purpose to describe the ideals H_i . They are unique in a strong sense in that if $R = H' \oplus H''$ where H' and H'' are ideals then H' (and H'') is a direct sum of a subset of $\{H_1,\ldots,H_n\}$.

Suppose that e and f are minimal idempotents. Then e and f are called <u>linked</u> if there is a sequence $e = e_1, e_2, \ldots, e_{n-1}, e_n = f$ of minimal idempotents such that $e_i R e_{i+1} \neq 0$; equivalently, the indecomposable left ideals Re_i and Re_j have a chain factor in common. The linking of minimal idempotents is an equivalence relation. The sum of <u>all</u> the indecomposable left ideals Re where e ranges over an equivalence class is called a <u>block</u> of R .

The above is nicely illustrated by appealing to the representation of R as a checkered matrix ring. Suppose that by lifting minimal idempotents from R/Rad(R) we have

$$1 = (\sum_{i=1}^{s} \sum_{j=1}^{s_i} e_i^{(j)}) + (\sum_{i=1}^{t} \sum_{j=1}^{t_i} f_i^{(j)}) + \cdots + (\sum_{i=1}^{u} \sum_{j=1}^{u_i} g_i^{(j)})$$

(analogous to the proof of (X.1)) where $\{e_i^{(j)}\},\{f_i^{(j)}\},\ldots,\{g_i^{(j)}\}$ are equivalence classes of linked idempotents. Proceed as in the proof of (X.1), setting $e_i = e_i^{(1)}, f_j = f_j^{(1)}, \ldots, g_k = g_k^{(1)}$ for $i,j,\ldots,k = 1,2,\ldots$.

Then there is a natural anti-isomorphism between R and the ring of matrices of the form

$$\begin{bmatrix} E & 0 & \cdots & 0 \\ 0 & F & \cdots & 0 \\ \vdots & \vdots & \ddots & \vdots \\ 0 & 0 & \cdots & G \end{bmatrix}$$

where E is a matrix

$$E = \begin{bmatrix} E_{11} & E_{12} & \cdots & E_{1s} \\ E_{21} & E_{22} & \cdots & E_{2s} \\ \vdots & \vdots & & \vdots \\ E_{s1} & E_{s2} & \cdots & E_{ss} \end{bmatrix}$$

with E_{ij} a s_i by s_j matrix over e_iRe_j . It is obvious the blocks

E,F,...,G determine (by the anti-isomorphism) ideals of R and that R is a direct sum of these ideals.

<u>(XI.3)</u> <u>THEOREM</u>. The blocks of a ring R are indecomposable two-sided ideals and R is their direct sum. Further, if $R = H_1 \oplus \cdots \oplus H_n$ is any decomposition of R into indecomposable two-sided ideals then the direct summands H_i are precisely the blocks of R.

<u>Proof</u>. It is only necessary to show the blocks are indecomposable and unique in the stated manner. Suppose that $R = H_1 \oplus \cdots \oplus H_n$ where the H_i are indecomposable two-sided ideals. Observe such a decomposition is always possible. We claim that the set $\{H_i\}$ is precisely the set of blocks of R.

Suppose e is a minimal idempotent. Then $Re = H_1e \oplus \cdots \oplus H_ne$. But Re is indecomposable so $Re = H_ie$ for some i. We note that ReR belongs to the same summand as Re. Then, if $eRf \neq 0$ (f minimal) we have $ReR \cap RfR \neq 0$ and thus RfR belongs to the same summand as ReR.

Now suppose e is linked to f. Then there is a sequence $e = e_1, e_2, \ldots, e_n = f$ with $e_iRe_{i+1} \neq 0$. Thus, by the above ReR and RfR belong to the same summand H_i. Thus, H_i is a direct sum of blocks. But the blocks are ideals and H_i is indecomposable. Thus H_i is a block of R.

The following exercises are from Michler [13].

(XI.1) EXERCISE. Show that a two-sided ideal H of a ring R is a block of R if and only if $H = Re$ where e is a central minimal idempotent of R.

(XI.2) EXERCISE. Let e be a central idempotent of R, i.e., e lies in the center of R. Suppose e is in the ideal H. Show the decompositions of e into central minimal idempotents of H and of R are the same.

(XI.3) EXERCISE. Let e and f be minimal idempotents of a ring R and $N = \mathrm{Rad}(R)$. Show the following are equivalent.

(a) e and f are linked.

(b) Re/Ne and Rf/Nf belong to the same block H of R.

(XI.4) EXERCISE. Let M be a finitely generated left R-module, e be a minimal idempotent of R and $N = \mathrm{Rad}(R)$.

(a) Show that M has a composition chain factor isomorphic to Re/Ne if and only if $eM \neq 0$. [Hint: See (XI.1) and (XII.2).]

(b) Show that M is a simple R-module if and only if M is R-isomorphic to Re/Ne for some minimal idempotent e of R. [Hint: See (XII.3) in the next section.]

(c) Two simple R-modules M_1 and M_2 are said to be <u>linked</u> if

M_1 and M_2 belong (via part (b) above) to the same block of R. Show there exist natural bijections between the following sets.

(1) The classes of linked simple left R-modules.

(2) The block ideals of R.

(3) The central minimal idempotents of R.

(For an interesting ring theoretic treatment of blocks and centers of group algebras over finite fields see Michler [13].)

XII. MODULES OVER A FINITE RING

In our presentation we have purposely selected an ideal theoretic approach to the structure of a finite ring. In a sense it is an internal approach to the theory of a finite ring and, by current standards, very classical. Presently it is fashionable to characterize a ring or obtain information about the ring in terms of the properties enjoyed by some class of modules over the ring. For example, it is easy to show that a finite ring is a field if and only if all its modules are free. We chose however the classical internal approach of examination of structure by way of left ideals. This choice was made because it was felt that the value of finite rings lies not necessarily in the study of the ring and its modules but rather in applications of finite rings to other research areas; e.g., combinatorics. In this setting it is generally more valuable to understand the ring than the module. In this section we use the results we have developed to examine modules over finite rings.

We initially examine the classification of simple modules and projective modules over finite rings. Modules over semi-simple rings are then described. This is followed with standard facts on Loewy chains, Cartan invariants, projective covers and injective modules. We conclude with some comments on indecomposable modules and modules over semi-local rings.

NOTE. Throughout this section we understand that an "R-module M " means a finitely generated left R-module.[1] The reader may wish to refer again to (III.).

An R-module M is said to be:

(a) indecomposable if $M \neq 0$ and M is not a direct sum of non-trivial R-modules.

(b) simple if M has no submodules other than M and 0 and $M \neq$

(c) completely reducible if every submodule of M is an R-direct summand of M .

---

[1] This hypothesis may often be omitted but the finitely generated case is the only case we wish to consider.

It is easy to see that an R-module M may be written as a direct sum

$$M = M_1 \oplus \cdots \oplus M_m$$

where the M_i are indecomposable R-modules. Further, if

$$M = N_1 \oplus \cdots \oplus N_n$$

is a second decomposition of M into indecomposable R-modules then the Krull-Schmidt Theorem [see (III.)] implies that $m = n$ and there is a bijection σ of $\{1,\ldots,m\}$ so that $M_i \simeq N_{\sigma(i)}$ for $1 \le i \le m$.

A descending chain [see (III.)]

$$M = M_1 \supset M_2 \supset \cdots \supset M_m \supset M_{m+1} = 0$$

of submodules of M is called a Jordan-Hölder chain if M_i/M_{i+1} is a simple non-zero R-module for $1 \le i \le m$. If

$$M = N_1 \supset N_2 \supset \cdots \supset N_n \supset N_{n+1} = 0$$

is a second Jordan-Hölder chain for M then $n = m$ and there is a bijection σ of $\{1,\ldots,m\}$ so that $M_i/M_{i+1} \simeq N_{\sigma(i)}/N_{\sigma(i)+1}$ for $1 \le i \le m$. The quotient modules M_i/M_{i+1} are called chain factors and the integer m is called the length of the chain.

(XII.1) THEOREM. The following are equivalent for a finitely generated R-module M .

(a) M is completely reducible.

(b) M is a direct sum of simple R-modules.

(c) M is a sum (not necessarily direct) of simple R-modules.

Proof. Clearly (a) implies (b) and (b) implies (c). It is only necessary to show (c) implies (a). Let N be a submodule of M. Select a submodule P which is maximal with respect to $P \cap N = 0$. We claim $P \oplus N = M$.

Suppose that $N + P \neq M$. Then there is an m in M with m not in $N + P$. By (c) $m = m_1 + \cdots + m_t$ where the m_i are in simple modules M_i. Since m is not in $N + P$ then, for some i, m_i is not in $N + P$. But then $M_i \cap (P + N) = 0$. Thus, $(M_i + P) \cap N = 0$. Since $M_i + P$ contains P properly this contradicts the choice of P. Thus $P + N = M$ and $M = N \oplus P$.

Utilizing a proof analogous to (XI.1) one can show the following.

(XII.2) THEOREM. Let M be a finitely generated R-module and e a minimal idempotent of R. Then M has a chain factor R-isomorphic to $Re/(\text{Rad}(R))e$ if and only if $eM \neq 0$.

(XII.3) COROLLARY. (Classification of Simple Modules). An R-module M is simple if and only if M is R-isomorphic to $Re/(\text{Rad}(R))e$ for some minimal idempotent e of R.

We note that (XII.3) permits the description of simple modules for all finite rings. We first observe that the action of R on $Re/(Rad(R))e$ (as an R-module) is the same as that of $R/Rad(R)$. For if $r \equiv r_1 \bmod(Rad(R))$ then $r = r_1 + s$ where s is in $Rad(R)$. Then for m in Re , $rm = r_1m + sm \equiv r_1m \bmod Rad(R)e$. We have the following simple modules.

(a) If R is local then its residue field k is its only simple module.

(b) If R is basic and $R/Rad(R) = k_1 \oplus \cdots \oplus k_t$ then the finite fields $k_1, \ldots, k_t$ are the only simple R-modules.

(c) If $R/Rad(R) = \oplus \sum_{i=1}^{t} (k_i)_{n_i}$ then the k_i-vector spaces $k_i^{(n_i)}$, $1 \le i \le t$, are the simple modules of R .

It is also clear that if R has the finite field k (resp., $k_1, \ldots, k_t$) as its only simple R-module then R is local (resp., basic).

The classification of all indecomposable modules is extremely difficult and is discussed later. The next step is thus to either condition the module or the ring. The following is available concerning projective modules.

<u>(XII.4)</u> <u>THEOREM</u>. (Classification of Projective Modules). A finitely generated R-module M is projective if and only if there exist minimal idempotents $\{e_1, \ldots, e_t\}$ in R with $M \simeq Re_1 \oplus \cdots \oplus Re_t$.

Proof. Since M is finitely generated and projective, M is a direct summand of a finitely generated free R-module F. But each direct summand ${}_R R$ of F is a direct sum of indecomposable left ideals of the form Re. Since M is a direct summand of F we may apply the Krull-Schmidt Theorem and deduce that $M \simeq Re_1 \oplus \cdots \oplus Re_t$ for some collection of minimal idempotents $e_1, \ldots, e_t$. The argument is reversible.

On the other hand, it is possible to classify all modules over a semi-simple ring.

(XII.5) THEOREM. (Classification of Modules over Semi-Simple Rings). The following are equivalent:

(a) R is a semi-simple ring.

(b) Every finitely generated R-module is a direct sum of simple R-modules.

(c) Every finitely generated R-module is projective.

Proof. If R is semi-simple, let $\{e_1, \ldots, e_n\}$ be a complete set of minimal orthogonal idempotents. Then $1 = \sum_{i=1}^{n} e_i$ and if M is an R-module

$$M = \sum_{m \in M} \sum_{i=1}^{n} Re_i m \qquad \text{(not necessarily direct).}$$

The map $Re_i \to Re_i m$ by $re_i \to re_i m$ is an R-morphism and since Re_i is simple it has kernel either 0 or Re_i. Thus M is a sum of simple modules and by (XII.1) a direct sum of simple modules.

If every R-module is a direct sum of simple R-modules then by (XII.1) every R-module is completely reducible. Hence R is completely reducible and every left ideal is a direct summand. The result follows by (VIII.5).

Hence (a) and (b) are equivalent. The remaining is a straightforward application of (XII.4).

We now examine the elementary facts concerning modules over rings with non-zero radicals.

Let M be an R-module. A chain

$$M = M_0 \supset M_1 \supset \cdots \supset M_{m-1} \supset M_m = 0$$

is called a Loewy chain if each quotient module M_i/M_{i+1} , $0 \le i \le m - 1$, is a direct sum of simple R-modules. Jordan-Hölder chains are Loewy chains of maximal length.

(XII.6) THEOREM. Let N be a submodule of a finitely generated R-module M . The quotient module M/N is completely reducible if and only if $\text{Rad}(R)M \subseteq N$.

Proof. If $\text{Rad}(R)M \subseteq N$ then M/N is naturally a $R/\text{Rad}(R)$ -module; i.e., the action of R on M/N is the action of $R/\text{Rad}(R)$. Thus M/N is completely reducible.

Conversely, if M/N is completely reducible then

$$M/N = M_1 \oplus \cdots \oplus M_m$$

where the M_i are simple non-zero modules. Suppose $\mathrm{Rad}(R)M \not\subseteq N$. Then $\mathrm{Rad}(R)M_i \neq 0$ for some i. Since M_i is simple, $\mathrm{Rad}(R)M_i = M_i$. By Nakayama's Lemma (V.2) $M_i = 0$, a contradiction.

The <u>Upper Loewy Chain</u> for M is constructed by selecting M_1 from among the submodules of M such that M/M_1 is maximal cardinality among the set of completely reducible quotient modules of M. Then a submodule M_2 of M_1 is chosen so that M_1/M_2 is maximal among the set of completely reducible quotient modules of M_1. Continue in this manner. The <u>Lower Loewy Chain</u> is formed by selecting in M a submodule N_1 which is maximal among the completely reducible submodules of M. Then choose N_2, $M \supset N_2 \supset N_1$, so that N_2/N_1 is a maximal completely reducible quotient module. Continue in this manner.

<u>(XII.7)</u> <u>THEOREM</u>. Let M be a finitely generated R-module and suppose β is such that $\mathrm{Rad}(R)^{\beta-1}M \neq 0$ and $\mathrm{Rad}(R)^{\beta}M = 0$.

(a) The chain

$$M \supset \mathrm{Rad}(R)M \supset (\mathrm{Rad}(R))^2M \supset \cdots \supset (\mathrm{Rad}(R))^{\beta}M = 0$$

is the upper Loewy chain of M.

(b) Let $N_t(M)$ be the submodule of M which is annihilated by $\mathrm{Rad}(R)^t$ but not by $\mathrm{Rad}(R)^{t-1}$. Then

$$M \supset N_{\beta-1}(M) \supset \cdots \supset N_1(M) \supset 0$$

is the lower Loewy chain of M .

(c) The upper and lower Loewy chains are unique and have the same length.

<u>Proof</u>. Parts (a) and (b) follow from (XII.6). To exhibit uniqueness for the upper Loewy chain suppose we have selected M_1 and M_1' so that M/M_1 and M/M_1' are maximal among completely reducible quotient modules. Then $M/(M_1 \cap M_1')$ is completely reducible. By (XII.6) $\mathrm{Rad}(R)M \subseteq M_1$ and $\mathrm{Rad}(R)M \subseteq M_1'$. Thus $\mathrm{Rad}(R)M \subseteq M_1 \cap M_1'$. But $M/M_1 \subseteq M/(M_1 \cap M_1')$ and since M/M_1 is maximal we must have $M/M_1 = M/(M_1 \cap M_2')$. Hence $M_1 \subseteq M_1'$. By symmetry $M_1' \subseteq M_1$ and we are done. The remaining assertation is straightforward.

<u>(XII.8)</u> <u>COROLLARY</u>. Let β denote the nilpotency of Rad(R) .

(a) If e is an idempotent of R , then

$$Re \supset \mathrm{Rad}(R)e \supset \cdots \supset (\mathrm{Rad}(R))^{\beta-1}e \supset 0$$

is the upper Loewy chain for Re .

(b) $R \supset \mathrm{Rad}(R) \supset \cdots \supset (\mathrm{Rad}(R))^{\beta-1} \supset 0$

is the upper Loewy chain for R .

The next topic of this section concerning in general modules over a ring of non-zero radical is a description of the "Cartan" invariants of a finite ring.

Let M be an R-module, e a minimal idempotent of R and

$$(*) \qquad M = M_1 \supset \cdots \supset M_m \supset M_{m+1} = 0$$

be a Jordan-Hölder chain for M. Then (XII.2) $eM \neq 0$ if and only if $M_i/M_{i+1} \simeq Re/(Rad(R))e$ (as R-modules) for some i.

First suppose $M_i/M_{i+1} \simeq Re/(Rad(R))e$. Then there is an element m_i in M_i with em_i not in M_{i+1}. We may assume $m_i = em_i$. Then m_i is not in eM_{i+1}. Then since M_i/M_{i+1} is simple and $0 \not\subseteq Rem_i \subset M_i$ we have that $Rem_i/M_{i+1} = M_i/M_{i+1}$ and $eRem_i/eM_{i+1} = eM_i/eM_{i+1}$.

If, on the other hand, $Re/(Rad(R))e$ is not isomorphic to M_i/M_{i+1} then $eM_i \subset M_{i+1}$. Thus $eM_i \subset eM_{i+1}$ and, since $eM_i \supset eM_{i+1}$, $eM_i = eM_{i+1}$.

In (*) select $m_{i_1},\ldots,m_{i_t}$ for which $Re/(Rad(R))e \simeq M_{i_j}/M_{i_j+1}$. Then

$$eM = eRem_{i_1} + \cdots + eRem_{i_t}$$

since $eRem_{i_j} \equiv eM_{i_j} \bmod eM_{i_j+1}$ and $eM_k = eM_{k+1}$ for $k \neq i_j$, $1 \leq j \leq t$.

<u>(XII.9)</u> <u>THEOREM</u>. Let M be a finitely generated R-module and

$$M = M_1 \supset \cdots \supset M_m \supset M_{m+1} = 0$$

be a Jordan-Hölder chain for M. Let e be a minimal idempotent of R. If t denotes the number of chain factors R-isomorphic to $Re/Rad(R)e$

then there exist t elements $\{m_1,\ldots,m_t\}$ in eM with

$$eM = eRem_1 + \cdots + eRem_t .$$

If the local ring $S = eRe$ has characteristic a prime then [(VIII.28)(c)] $S = k + eRad(R)e$ where $k = S/eRad(R)e$. In this case, $Rad(R)em_i/M_{i+1}$ is a submodule of M_i/M_{i+1} . Since M_i/M_{i+1} is simple $Rad(R)em_i/M_{i+1} = 0$ and thus $eRad(R)em_i/M_{i+1} = 0$. Hence $km_i/eM_{i+1} = eM_i/eM_{i+1}$ and $eM = km_1 + \cdots + km_t$ is a k-vector space of dimension t .

We apply the above discussion now to the R-module ${}_RR$. Suppose ${}_RR$ has a decomposition

$${}_RR = Re_1 \oplus \cdots \oplus Re_n$$

into indecomposable left ideals. We consider the upper Loewy series of Re_i , namely,

$$Re_i \supset Rad(R)e_i \supset \cdots \supset Rad(R)^{\beta-1}e_i \supset 0$$

and the completely reducible quotient module

$$N_{ij} = (Rad(R)^j)e_i/(Rad(R)^{j+1})e_i .$$

<u>(XII.10)</u> <u>THEOREM</u>. (Same notation as above.)

(a) The number of simple submodules of N_{ij} which are R-isomorphic to the simple module $Re_k/\mathrm{Rad}(R)e_k$ is the length $l_{ij}^{(k)}$ of the Jordan-Hölder chain of the e_kRe_k-module e_kN_{ij} .

(b) The number of chain factors of Re_i isomorphic to $Re_k/\mathrm{Rad}(R)e_k$ is equal to the length L_{ik} of the Jordan-Hölder chain of e_kRe_i as an e_kRe_k-module. Further

$$L_{ik} = \sum_{j=0}^{t-1} l_{ij}^{(k)}$$

where t is the length of the Loewy series of Re_i .

If $Re_i \simeq Re_s$ and $Re_j \simeq Re_t$ then $L_{ij} = L_{st}$. Let $C_1,\ldots,C_m$ denote the distinct classes of isomorphic minimal idempotents. Then the integers L_{ij} belong to distinct pairs of classes are called the Cartan Invariants.

Theorem (XII.4) provided a description of projective modules. However, (XII.4) indicates that if the ring is not semi-simple then the ring possesses modules which are neither free nor projective. We would like to show now that if M is an R-module then it is possible to find a projective R-module P and a surjective R-morphism $\sigma : P \to M$ such that the kernel of σ is "small" in P . In a sense, M is "nearly" projective.

If M is an R-module, a submodule N of M is called small if whenever we have a submodule Q of M with $M = Q + N$ then $M = Q$. Obviously 0 is a small submodule of M . A projective cover of a module M is a projective R-module P and a surjective R-morphism $\sigma : P \to M$ such that $\ker(\sigma)$ is small in P. Note that if $\ker(\sigma) = 0$ then $P \simeq M$.

(XII.11) THEOREM. Let M be an R-module. Then M has a projective cover. Further, this cover is unique up to R-module isomorphism.

Proof. (Bass). Clearly $M/\mathrm{Rad}(R)M \simeq \oplus \sum_{i=1}^{n} Re_i/\mathrm{Rad}(R)e_i$, where e_i is an idempotent in R. Take $P = \oplus \sum_{i=1}^{n} Re_i$. We have

$$\begin{array}{c} P \\ \downarrow \\ M \to M/\mathrm{Rad}(R)M \to 0 \end{array} \qquad \text{(exact)}.$$

Thus there is a map $\sigma : P \to M$ such that the diagram commutes. Since P is projective, it remains only to show that σ is surjective and $\ker(\sigma)$ is small.[1]

First observe that $\mathrm{Im}(\sigma) + \mathrm{Rad}(R)M = M$. Thus by Nakayama's Lemma (V.2) $\mathrm{Im}(\sigma) = M$ and σ is surjective.

Next note that $\ker(\sigma) \subset \ker(P \to M/\mathrm{Rad}(R)M) = \mathrm{Rad}(R)P$. Thus, for any submodule Q of P if $\ker(\sigma) + Q = P$ then $\mathrm{Rad}(P)P + Q = P$ and again by Nakayama's Lemma $Q = P$. Hence $\ker(\sigma)$ is small.

---

[1] This proof gives the construction of a projective cover.

It remains to show uniqueness. Let $\sigma : P \to M$ and $\sigma' : P' \to M$ be two projective covers for M . Then there is a ϕ so that

$$\begin{array}{ccccc} & & P' & & \\ & \swarrow^{\phi} & \downarrow \sigma' & & \\ P & \xrightarrow{\sigma} & M & \longrightarrow & 0 \end{array}$$

is commutative; i.e., $\sigma\phi = \sigma'$. Since σ and σ' have small kernels ϕ is surjective.

Thus, we have

$$0 \longrightarrow \ker \phi \longrightarrow P' \xrightarrow{\phi} P \longrightarrow 0$$

exact and hence the sequence splits. That is, we have $\rho : P \to P'$ so that $\phi\rho = id_P$. The diagram

$$\begin{array}{ccccc} P & & & & \\ \rho\downarrow & \searrow^{id_P} & & & \\ P' & \xrightarrow{\phi} & P & \xrightarrow{\sigma} & M \end{array} \qquad \sigma' = \sigma\phi : P' \to M$$

shows that ρ is surjective. Thus $P \simeq P'$.

Consider the following situation. Suppose M, P and Q are R-modules with P as a submodule of Q . Suppose that we have an R-morphism $\sigma : P \to M$. It is often useful to "lift" this morphism to an R-morphism $\sigma' : Q \to M$ such that $\sigma'|_P = \sigma$, i.e., it is useful to extend σ to all of Q . Unfortunately this is not always possible. We now define a class of modules possessing the property of M .

A left R-module M is called *injective* or *R-injective* if whenever we have the following diagram of modules and morphisms

$$\begin{array}{ccccc} & & M & & \\ & & \uparrow \rho & & \\ Q & \xleftarrow{\beta} & P & \longleftarrow & 0 \end{array} \qquad \text{(row exact)}$$

there exists an R-morphism $\rho^* : Q \to M$ such that $\rho^*\beta = \rho$.

(XII.12) *THEOREM*. Let M be a finitely generated R-module. The following are equivalent:

(a) M is injective

(b) Every short exact sequence $0 \to M \to N$ of R-modules splits. Thus, M is a direct summand of every module which contains it.

(c) For every left ideal L of R and R-morphism $\sigma : L \to M$ there is an R-morphism $\sigma' : R \to M$ which extends σ .

(d) For every left ideal L of R and R-morphism $\sigma : L \to M$ there is an x in M with $\sigma(a) = ax$ for all a in L .

(e) If P is a submodule of Q and $\sigma : P \to M$ is an R-morphism then there is an R-morphism $\sigma' : Q \to M$ extending σ

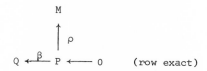

Proof. We show only that (c) implies (e). The other implications are straightforward. Assume (c) and suppose P is a submodule of Q. Suppose further that $\sigma : P \to M$ is an R-morphism. If $P = Q$ we are done. Thus, suppose $P \neq Q$ and x is in Q but not P. Take $L = \{a \text{ in } R \mid ax \text{ is in } P\}$. Observe that L is a non-empty left ideal of R.

Define $\beta : L \to M$ by $\beta(a) = \sigma(ax)$. By (c) there is an R-morphism $\beta' : R \to M$ extending β. Let $y = \beta'(1)$. Then for ax in P, $\sigma(ax) = \beta(a) = \beta'(a) = ay$.

Let $P' = P + Rx$ and define $\sigma' : P' \to M$ by $\sigma'(p + ax) = \sigma(p) + ay$. It is easy to check that this map is well-defined and $\sigma'|_P = \sigma$.

If $P' = Q$ we are done. If not, repeat the argument with P' replacing P. Since Q is finite we eventually extend σ to all of Q.

<u>(XII.13)</u> <u>THEOREM</u>. (Continuation of (XII.5).) The following are equivalent:

(a) R is a semi-simple ring.

(b) Every finitely generated R-module is injective.

<u>Proof</u>. Suppose that R is semi-simple and $0 \to M \xrightarrow{\sigma} N$ is an exact sequence of R-modules. By (XII.5) and (XII.1), $\sigma(M)$ is a direct summand of N. Thus the sequence splits and M is injective.

Conversely, let M be an R-module and N a submodule of M . Since N is injective, it is a direct summand of M . Thus M is completely reducible. The result again follows from (XII.5) and (XII.1).

The above result indicates that injective modules are plentiful.

We know that every R-module M is a direct sum of indecomposable modules. If the indecomposable modules are simple then M is completely reducible and the simple summands are described following (XII.3). The principal indecomposable modules are modules of the form Re where e is a minimal idempotent of R . It is clear that ${}_R R$ decomposes as a direct sum of principal indecomposable modules and by (XII.3) every simple module is isomorphic to Re/(Rad(R))e for some principal indecomposable module. Further, every projective module is a direct sum of principal indecomposable modules (XII.4). However, as noted before (XII.4), in general the classification of the indecomposable finitely generated modules over a ring is a difficult unsolved problem. We now provide some examples to illustrate this. Indeed, it is not even true that for a given ring there exist only a finite number of isomorphism classes of finitely generated indecomposable modules.

The first example is due to Brenner [23]. For a ring R , let $T_2(R)$ denote the ring of 2×2 lower triangular matrices over R . Observe that $T_2(R)$ is generated over R by the matrix units E_{11} , E_{21} , E_{22} . And $T_2(R)$-module M is equivalent to a pair

$$M_1 = E_{11}M \qquad M_2 = E_{22}M$$

of R-modules and a map $\sigma : M_1 \to M_2$ induced by E_{21} . Conversely, given any two R-modules M_1 and M_2 and an R-morphism $\sigma : M_1 \to M_2$, it is possible to construct an equivalent $T_2(R)$-module which we denote by (M_1,M_2,σ) . Let $T_2 = T_2(R)$. If $M = (M_1,M_2,\sigma)$, then $\mathrm{End}_{T_2}(M)$ consist of those elements of $\mathrm{End}_R(M)$ which commute with E_{11},E_{21} and E_{22} . We conclude that $\mathrm{End}_{T_2}(M)$ is the subring of $\mathrm{End}_R(M_1) \oplus \mathrm{End}_R(M_2)$ of elements $\langle\alpha,\beta\rangle$ satisfying $\sigma\alpha = \beta\sigma$.

<u>(XII.14)</u> <u>THEOREM</u>. (Brenner [23]). If $t \geq 4$ and n is a positive integer, then there is an indecomposable $T_2(Z/Zp^t)$-module (p a prime) with minimal generating set of $3n$ elements.

<u>Proof</u>. (Brenner). Let n be a positive integer and A a free finitely generated Z/Zp^4-module with a minimal generating set of n elements. Let $B = p^2A$ and $M_1 = M_2 = A \oplus B$. Any morphism ϕ in $\mathrm{Hom}_{Z/Zp^4}(M_i,M_j)$ can be represented as

$$\begin{bmatrix} \phi_{11} & \phi_{12} \\ p^2\phi_{21} & \phi_{22} \end{bmatrix}$$

where ϕ_{ij} is in $\mathrm{End}(A)$.

Let J be a cyclic endomorphism of B and $\sigma : M_1 \to M_2$ be given by

$$\begin{bmatrix} pI & I \\ p^2 I & pJ \end{bmatrix}$$

where I is the identity morphism.

If

$$\alpha = \begin{bmatrix} \alpha_{11} & \alpha_{12} \\ p^2\alpha_{21} & \alpha_{22} \end{bmatrix} \quad \text{and} \quad \beta = \begin{bmatrix} \beta_{11} & \beta_{12} \\ p^2\beta_{21} & \beta_{22} \end{bmatrix}$$

are morphisms of M_1 and M_2 , respectively, and

$$\sigma\alpha = \beta\sigma ,$$

then $\alpha_{11} = \alpha_{22} = \beta_{11} = \beta_{22}$ (mod p) and $\alpha_{11}J = J\alpha_{11}$ (mod p) .

Since J is cyclic, then modulo p , the morphism α_{11} is a polynomial in J .

By the discussion prior to the theorem we thus have that $\mathrm{End}_{T_2(Z/Zp^4)}(M_1,M_2,\sigma)$ consists of all $\langle\alpha,\beta\rangle$ satisfying the above. One checks that this ring contains no proper idempotents. Hence it is local and (M_1,M_2,σ) is indecomposable. Finally, examination of M_1 , M_2 and σ indicates that $\sigma M_1 = pA \oplus B$. Thus the cardinality of a minimal generating set of (M_1, M_2, σ) is $3n$.

<u>(XII.15)</u> <u>THEOREM</u>. (Heller and Reiner [33]). Let $G = \langle a \rangle \times \langle b \rangle$ be a direct product of two cyclic groups of order p (p a prime). Let k be a finite field of characteristic p . Then, if n is a positive integer $k[G]$ has an indecomposable module of k-dimension $2n + 1$.

<u>Proof</u>. (Heller and Reiner). Let M be a k-vector space with k-basis $\{x_0, x_1, \ldots, x_n; y_1, \ldots, y_n\}$. Define

$$ax_i = bx_i = x_i \ , \quad 0 \le i \le n$$
$$(a - 1)y_i = x_i \ , \quad 1 \le i \le n$$
$$(b - 1)y_i = x_{i-1} \ , \quad 1 \le i \le n \ .$$

Then M is a $k[G]$-module. Let $M = N \oplus P$ as a k-space with N spanned by $\{x_i\}$ and P by $\{y_j\}$.

Left multiplications by $a - 1$ and $b - 1$ map P into N and map N to 0 .

Suppose that, as a $k[G]$-module, M is decomposable. Let

$$M = M_1 \oplus M_2 \ .$$

Let $\Pi : M \to N$ be the natural projection and suppose that $\dim_k \Pi M_1 = r$ for some r , $1 \le r \le n$. Observe $(a - 1)M_1 \subset M_1$ and $(a - 1)M_1 = (a - 1)\Pi M_1$. Thus $\dim_k (a - 1)M_1 = r$. Since $(b - 1)M_1$ contains at least one vector not in $(a - 1)M_1$ we have $\dim_k (b - 1)M_1 \ge r + 1$

and $\dim_k M_1 \geq 2r + 1$. An analogous result is true for M_2 ; i.e., $\dim_k M_2 \geq 2s + 1$ for some s , $1 \leq s \leq n$. Then $r + s \geq n$ and

$$\dim_k M \geq 2(r + s) + 2 \geq 2n + 2$$

- a contradiction. Thus M is indecomposable.

It was shown in VIII (precisely (VIII.27)) that a ring R has a decomposition

$$R = (\oplus \sum_{i=1}^{n} M_{m_i}(S_i)) \oplus N$$

as an Abelian group where N is an additive subgroup of $Rad(R)$ and the S_i , $1 \leq i \leq n$, are local rings. It has been noted that in various cases $N = 0$. If $N = 0$, then as a ring R is a direct sum of matrix rings over local rings. The final part of this section is concerned with the examination of R when $N = 0$ and its modules. If the S_i are finite fields and $N = 0$ then R is semi-simple and its modules are characterized by (XII.5) and (XII.13).

Recall a ring R is <u>semi-local</u> if $R/Rad(R)$ is a matrix ring over a finite field. By (VIII.26) a semi-local ring is isomorphic to an n by n matrix ring over a local ring. The integer n is unique and the local ring is unique up to isomorphism.

A ring R is called <u>projectively trivial</u> if there is a projective R-module P , called a <u>minimal projective</u> of R , such that every finitely generated projective R-module M is of the form $\oplus \sum_{i=1}^{n} P$ for some unique integer n determined by M .

<u>(XII.16)</u> <u>THEOREM</u>. A ring R is projectively trivial if and only if R is semi-local.

<u>Proof</u>. Let $R = M_n(S)$ and E_{11} be the standard matrix unit. If e is a minimal idempotent of R then $Re \simeq RE_{11}$. Take $P = RE_{11}$ and apply (XII.4). The converse also follows from (XII.4). For, if $_RR = \oplus \sum_{i=1}^{n} P$ (P minimal projective), then $P \simeq Re$ where e is a minimal idempotent and one applies (VII.3).

The above result characterizes a semi-local ring in the context of its projective modules. It has already been observed that R is a finite field if and only if every R-module is free, and R is semi-simple if and only if every R-module is projective. Recently Courter [24] gave necessary and sufficient conditions on the modules of a ring R to assure that the ring was a direct sum of matrix rings over local rings. We provide a portion of Courter's results.

Let M be a finitely generated R-module. An arbitrary homomorphic image of M will be denoted by M/* . Then $\mathrm{Hom}_R(N,M/*) = 0$ will mean that $\mathrm{Hom}_R(N,\overline{M}) = 0$ <u>for every</u> homomorphic image $\overline{M}$ of M .

A finitely generated R-module M is *corationally complete* if whenever N is a finitely generated R-module with submodule V satisfying $\mathrm{Hom}_R(M,V/*) = 0$ then

$$\mathrm{Hom}_R(M,N) \to \mathrm{Hom}_R(M,N/V)$$

is surjective. That is, for each $\sigma : M \to N/V$ there is a $\sigma' : M \to N$ completing the following diagram

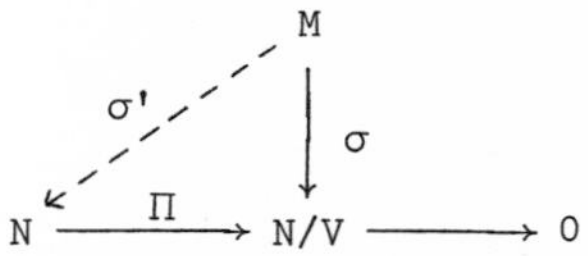

where Π is the natural projection.

Let M be a finitely generated R-module with submodule N. Then M is a *coessential extension of* N if N is small in M, i.e., whenever $P + N = M$ then $P = M$.

The finitely generated R-module M is a *corational extension* of a submodule N if $\mathrm{Hom}_R(M,N/*) = 0$, i.e. $\mathrm{Hom}_R(M,\overline{N}) = 0$ for every homomorphic image of N.

(XII.17) THEOREM. (Courter [24]). Let R be a ring. The following are equivalent.

(a) R is a direct sum of semi-local rings.

(b) Every finitely generated R-module is corationally complete.

(c) No finitely generated R-module is a corational extension of a proper submodule.

<u>Proof</u>. The proof will be long. The initial aim will be to show (b) implie (c).

Suppose an R-module M is a corational extension of a submodule N. Then, if M is not a coessential extension of N, there exists a submodule $P \neq M$ with $N + P = M$. Then $M/P \simeq N/N \cap P \neq 0$. Thus M is not a corational extension of N - a contradiction. Consequently, corational extensions are coessential extensions.

Assume that every R-module is corationally complete. Suppose that M is a corational extension of a submodule N. Thus $\mathrm{Hom}_R(M,N/*) = 0$. Clearl then

$$\mathrm{Hom}_R(M/N,N/*) = 0 .$$

Since M/N is corationally complete, there is a morphism $\sigma : M/N \to M$ such that

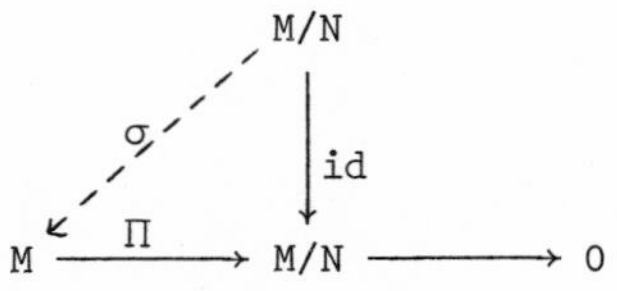

is a commutative diagram. Thus, the sequence $M \to M/N \to 0$ splits and $M = N \oplus \sigma(M/N)$. By the previous paragraph, N is small. Thus

$\sigma(M/N) = M$ and $N = 0$, i.e., N is not a proper submodule of M . This shows that (b) implies (c).

We now show that (c) implies (b). Assume (c) and suppose M is not corationally complete. Thus there is an R-module N with submodule V satisfying $\mathrm{Hom}_R(M,V/*) = 0$; however, for the diagram

(*)

$$\begin{array}{ccccc} & & M & & \\ & & \big\downarrow{\scriptstyle\sigma} & & \\ N & \xrightarrow{\Pi} & N/V & \longrightarrow & 0 \end{array}$$

there is no completing morphism $\sigma' : M \to N$. We show that $\mathrm{Hom}_R(M,V/*) = 0$ is not possible and obtain a contradiction.

We may assume without loss that σ in (*) is surjective. Let (P,β) be a projective cover for N/V with kernel $K = \ker(\beta)$. We have

$$\begin{array}{ccccc} & & M & & \\ & & \big\downarrow{\scriptstyle\sigma} & & \\ & & N/V & \longleftarrow & P \\ & & \big\downarrow{\scriptstyle id} & & \\ N & \xrightarrow{\Pi} & N/V & \longrightarrow & 0 \end{array} .$$

Then there is an α such that

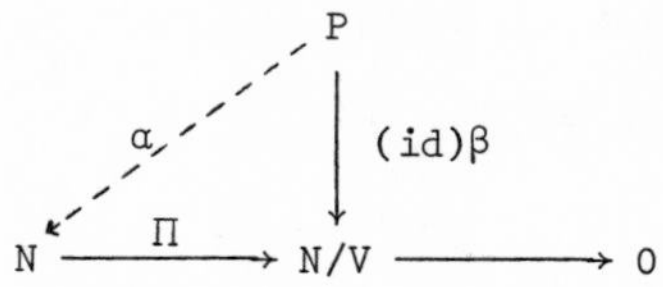

is commutative since P is projective. Let $L = \ker(\alpha)$. Then $K \cap L \subset K$ and $K \neq K \cap L$, otherwise, if $K = K \cap L$ then the above diagram provides a completion for (*). Thus $P/K \cap L$ is a proper extension of $K/K \cap L$. By hypothesis $P/K \cap L$ is not a corational extension. Thus a non-zero morphism ϕ of $\mathrm{Hom}_R(P/K \cap L,(K/(K \cap L))/*)$ exists. But K is small in P , so $K + \ker(\phi) \neq P$. Then ϕ induces a non-zero morphism on $P/K \simeq N/V$ and $\mathrm{Hom}_R(N/V,(K/(K \cap L))/*) \neq 0$.

Note that $K/K \cap L \simeq V \cap \alpha(P)$. Thus there is a non-zero morphism of N/V onto a quotient module, say $(V \cap \alpha(P))/Q$, of $V \cap \alpha(P)$. But Q is a submodule of V , hence $\mathrm{Hom}_R(M,V/*) \neq 0$.

We next show (a) implies (c). Assume R is a direct sum of semi-local rings. Let M be an R-module. We show initially that M has a maximal proper submodule. It is easy to see that if P is a projective R-module then P has a maximal submodule. Thus let $\sigma : P \to M$ with $K = \ker(\sigma)$ be a projective cover of M . If N is a maximal submodule of P then $K + N \neq P$ since K is small. Hence $K \subset N$ since N is maximal. Thus N/K gives a maximal submodule of $M \simeq P/K$.

Assume (c) is false. Suppose there is a module M with submodule $N \neq 0$ such that M is a corational extension of N. Thus $\mathrm{Hom}_R(M,N/*) = 0$.

Let $e_1,\ldots,e_n$ with $1 = e_1 + \cdots + e_n$ denote the central idempotents of R. For any left module Q, let Q_i denote e_iQ. Then $Q = \oplus \sum_{i=1}^{n} Q_i$.

Let β, $1 \leq \beta \leq n$, be such that $N_\beta \neq 0$. Let F' be a maximal submodule of M_β and let $F = F' + \sum_{i\neq\beta} M_i$. Let E' be a maximal submodule of N_β and $E = E' + \sum_{i\neq\beta} N_i$. Then $M/E \simeq N/F$. Thus M is not a corational extension of N.

We now complete the proof by showing (c) implies (a). It suffices to show that the central idempotents of $R/\mathrm{Rad}(R)$ lift to orthogonal idempotents which are central, i.e., commutative, in R. Thus let $\{e_1,\ldots,e_n\}$ be a set of orthogonal idempotents in R satisfying

$$1 = e_1 + \cdots + e_n$$

and

$$e_ir - re_i$$

in $\mathrm{Rad}(R)$. We claim the e_i are in the center of R. Suppose they are not. Then $re_i \neq e_ir$ for some r in R and some i, $1 \leq i \leq n$. Thus either $e_ire_i \neq e_ir$ or $e_ire_i \neq re_i$. Without loss we assume $e_ire_i \neq e_ir$. Consequently, for some j, $1 \leq j \leq n$, with $i \neq j$

$$y = e_ire_j \neq 0 .$$

Let $V = \sum_{t \neq j} Re_t$. Then $y = ye_j$ is not in V . Select H with $H \supseteq V$, H a left ideal of R , and H maximal with respect to exclusion of y . Let

$$F = \cap \{L \mid L \text{ left ideal, } H \not\subseteq L\} .$$

Then F/H is simple and $F = H + Ry$. Further, since R/H is a proper extension of R/F it cannot be corational. Hence there is a morphism

$$\phi : R/H \to R/F$$

with $\phi \neq 0$. Let $G = \ker(\phi)$. Then $R/G \simeq F/H$.

Thus, since $G \supseteq H \supseteq V$, $R \neq G$ and e_j is not in G we have

$$R = G + Re_j .$$

Further, note $e_j y = 0$.

But $e_j r - re_j$ is in $Rad(R)$ for all r in R . Since $(G + Re_j)/G$ is simple, $(G:e) = \{r \text{ in } R \mid re \text{ is in } G\}$ is a maximal left ideal and thus contains the intersection $Rad(R)$ of all the maximal left ideals. Since $e_j r - re_j$ is in $Rad(R)$, $e_j r - re_j$ is in $(G:e)$. Thus for each r in R , $e_j re_j - re_j$ is in G . Hence $re_j + G = e_j(re_j + G)$ and e_j is a left identity on $(Re_j + G)/G$. On the other hand, $y + H$ is not zero in $(Ry + H)/H$. But $e_j(y + H) = 0$ since $e_j y = 0$. Hence $(Re_j + G)/G \not\simeq (Ry + H)/H$ contradicting the above.

This completes the proof of (XII.17).

<u>(XII.1)</u> <u>EXERCISE</u>. For a ring R show that the following are equivalent.

(a) Every finitely generated projective R-module is free.

(b) R is a local ring.

(Hint: Use the fact that R is finite.)

<u>(XII.2)</u> <u>EXERCISE</u>. (Quasi-projective modules). (Golan, Wu, Jans). An R-module Q is called <u>quasi-projective</u> if whenever

$$\begin{array}{ccccc} & & Q & & \\ & & \downarrow{\scriptstyle \alpha} & & \\ Q & \xrightarrow{\beta} & M & \longrightarrow & 0 \end{array} \qquad \text{(row exact)},$$

there is an R-morphism $\alpha^* : Q \to Q$ such that $\beta\alpha^* = \alpha$.

(a) Let $0 \to K \to P \to Q \to 0$ be exact with P projective. If K is <u>stable</u>, i.e., $\sigma K \subseteq K$ for all σ in $\mathrm{End}_R(P)$, show that Q is quasi-projective.

(b) Let P be projective and Q be quasi-projective. Show that $P \to Q \to 0$ splits if $P \oplus Q$ is quasi-projective.

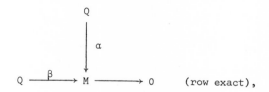

(c) Show that if $R \oplus M$ is quasi-projective for every simple module M then R is semi-simple.

(d) (Golan). Show the following are equivalent:

(1) R is semi-simple.

(2) Finite direct sums of quasi-projective modules are quasi-projective.

(3) Every module is quasi-projective.

(Thus, generally, direct sums of quasi-projective modules need not be quasi-projective.)

(e) If M is quasi-projective and $M = N \oplus P$, show N and P are quasi-projective.

(f) Let $\sigma : P \to M$ be surjective and suppose P is projective. Show M is projective if and only if $P \oplus M$ is quasi-projective

(g) If M is quasi-projective, show $\oplus \sum_{i=1}^{n} M$ is quasi-projective.

(h) If M is quasi-projective, show (use Krull-Schmidt Theorem) that M is a direct sum of indecomposable quasi-projective modules.

(i) Let M be quasi-projective and have projective cover $P(M) \xrightarrow{\Pi} M \longrightarrow 0$. Show that if $P = P_1 \oplus P_2$ then $M = M_1 \oplus M_2$ and $\Pi_i : P_i \to M_i \to 0$ is a projective cover of M_i where $\Pi_i = \Pi|_{P_i}$.

(j) Let M be quasi-projective. Show that M is indecomposable if and only if its projective cover is indecomposable.

(k) (Classification of quasi-projectives). (Wu, Jans). Let M be an indecomposable quasi-projective R-module. Then $M = Re/J \cap Re$

where e is a minimal idempotent of R and J is a two-sided ideal of R. (Hint: Use (j) and deduce $P(M) = Re$. Then $M = Re/L$ where L is a $End_R(Re)$-stable R-submodule of Re. Show $L = J \cap Re$ for J an ideal of R (take $J = Le + LeR(1 - e)$). Conversely, if J is an ideal show $J \cap Re$ is $End_R(Re)$-stable.)

(XII.3) EXERCISE. (See Notation of (XII.7)). Let M be a finitely generated R-module. Show $(Rad(R))^{i-1}M/(Rad(R))^{i}M$ has at least one simple summand in common with $N_{\beta-i+1}(M)/N_{\beta-i}(M)$.

(XII.4) EXERCISE. (Tensor Product). Let M_R and ${}_RN$ be finitely generated right and left R-modules. Form $M \times N = \{<m,n> \mid m \in M, n \in N\}$. Construct the free Z-module F consisting of all finite sums $\sum s<m,n>$ with s in Z. (If $\chi(R) = p^t$ (p a prime) then Z may be replaced by Z/Zp^t.) Let T be the subgroup generated by

$$<m + n,p> - <m,p> - <n,p>$$
$$<m,p + q> - <m,p> - <m,q>$$
$$<m,rp> - <mr,p>$$

where m,n are in M, p,q are in N and r is in R. Denote the quotient group F/T by $M \otimes_R N$ and denote the coset $<m,n> + T$ by $m \otimes n$.

(a) If S is a ring and M_R is a left S-module or ${}_RN$ is a right S-module, show $M \otimes_R N$ is naturally an S-module under

$s(\sum m_i \otimes n_i) = \sum (sm_i) \otimes n_i$ or $(\sum m_i \otimes n_i)s =$ $\sum m_i \otimes (n_i s)$, respectively.

(b) If N_1 and N_2 are left R-modules, show

$$M \otimes_R (N_1 \oplus N_2) \simeq (M \otimes_R N_1) \oplus (M \otimes_R N_2)$$

(as Abelian groups).

(c) If S is a ring and M and P are left S-modules, show

$$\text{Hom}_R(N,\text{Hom}_S(M,P)) \simeq \text{Hom}_S(M \otimes_R N,P) .$$

(d) Let $\sigma : N_1 \to N_2$ be an R-morphism for left R-modules N_1 and N_2 . Show $1 \otimes \sigma : M \otimes_R N_1 \to M \otimes_R N_2$ by $(1 \otimes \sigma)(\sum m \otimes n) =$ $\sum m \otimes \sigma(n)$ is a group morphism.

(e) Let $N_1 \to N_2 \to N_3 \to 0$ be an exact sequence of left R-modules. Show (using mapping described in (d)) that

$$M \otimes_R N_1 \to M \otimes_R N_2 \to M \otimes_R N_3 \to 0$$

is exact. Formulate and prove the analogous statement for $(\ ) \otimes_R N$.

(f) Let ${}_S M_R$ and ${}_R N_T$ be two-sided modules over rings R,S and T . Show $M \otimes_R N$ is T-projective if M is R-projective and N is T-projective.

(g) Show $R \otimes_R N \simeq N$ (as left R-modules).

(h) Assume R is a commutative ring.

(1) Show $M \otimes_R N$ is an R-module under $r(\sum m \otimes n) =$ $\sum (rm) \otimes n$. [Note: If R is commutative then any right

R-module M may also be regarded as a left R-module under $r \cdot m = mr$.]

(2) If P is an R-module, show

$$(M \otimes_R N) \otimes_R P \simeq M \otimes_R (N \otimes_R P)$$

and

$$M \otimes_R N \simeq N \otimes_R M$$

as R-modules.

(3) Let I and J be ideals of R . Show

$$(R/I) \otimes_R (R/J) \simeq R/(I + J)$$

REMARK. The above exercises provide a terse introduction to the tensor product of modules. Detailed information is available in most texts on ring theory and modern algebra.

(XII.5) EXERCISE. (Duality for Projective Modules). [See duality for matrix rings and basic rings.] Let R be a ring and M a finitely generated left R-module. Let $S = \mathrm{End}_R(M)^\circ$ where $\circ$ denotes the opposite ring and $M^* = \mathrm{Hom}_R(M,R)$. For σ in S write $m\sigma$ instead of $\sigma(m)$. Then ${}_RM_S$ is an (R,S)-bimodule, i.e., a left R-module, a right S-module and $(rm)\sigma = r(m\sigma)$. Similarly ${}_SM^*_R$ is an (S,R)-bimodule where if r is in R , m^* is in M^* and σ is in S , then $\sigma m^* r$ in M^* is given by (writing maps on right)

$$(x)\sigma m^* r = (((x)\sigma)m^*)r$$

for x in M .

Define an (R,R)-module morphism

$$(\, , \,) : M \otimes_S M^* \to R$$

by

$$m \otimes m^* \to (m,m^*) = m^*(m)$$

and an (S,S)-module morphism

$$[\, , \,] : M^* \otimes_R M \to S$$

by

$$(x)[m^*,m] = (x,m^*)m \ .$$

An R-module M is called a <u>generator</u> if R is a direct summand of copies of M . Further, M is called a <u>progenerator</u> or <u>faithfully projective</u> if M is a finitely generated projective generator.

<u>Assume M is a progenerator.</u>

(a) Show $(\, , \,) : M \otimes_S M^* \to R$ is surjective.

(b) Show $[\, , \,] : M^* \otimes_R M \to S$ is surjective.

(c) Show $(\, , \,)$ and $[\, , \,]$ are isomorphisms.

Let ${}_R\text{Mod}$ denote the class of finitely generated left R-modules and ${}_S\text{Mod}$ denote the class of finitely generated left S-modules.

(d) Show $M \otimes_S (\ ) : {}_S\text{Mod} \to {}_R\text{Mod}$ and $M^* \otimes_R (\ ) : {}_R\text{Mod} \to {}_S\text{Mod}$.

(e) For N in ${}_R\text{Mod}$ and T in ${}_S\text{Mod}$ show $M \otimes_S (M^* \otimes_R N) \simeq N$ and $M^* \otimes_R (M \otimes_S T) \simeq T$.

(f) Show M^* is a progenerator for ${}_S\text{Mod}$.

(g) Show $M \simeq \text{Hom}_S(M^*,S)$ and $R \simeq \text{End}_S(M^*,M^*)^{\circ}$.

(h) Show there exists a lattice isomorphism between the two-sided ideals of R and the two-sided ideals of S .

(For hints and additional results see Bass [4].)

(XII.6) EXERCISE. Let $\Pi : P \to M$ and $\sigma : Q \to N$ be projective covers. Show that $\Pi \oplus \sigma : P \oplus Q \to M \oplus N$ is a projective cover.

(XII.7) EXERCISE. Prove Theorem (XII.10).

(XII.8) EXERCISE. Complete the proof of Theorem (XII.12).

(XII.9) EXERCISE. (See (XII.14).) Show $T_2(Z/Zp^t)$ $(t \leq 3)$ has only a finite number of isomorphism classes of finitely generated indecomposable modules.

(XII.10) EXERCISE. (Injective Modules.)

(a) Show $\oplus \sum_{i=1}^{n} M_i$ is injective if and only if M_i is injective for each i .

(b) If M is an injective R-module, show that M is a direct sum of indecomposable injective R-modules.

(c) Let R be a local ring with a principal maximal ideal. Show that $_R R$ is an injective R-module.

(d) Show Z/nZ is an injective Z/nZ-module for an integer $n > 1$.

(e) For a finite field k , show k[X]/I is an injective k[X]/I-module for any ideal $I \neq 0$.

(XII.11) EXERCISE. (Self-injective rings.) A ring R is called self-injective or quasi-Frobenius if ${}_R R$ is an injective R-module. The previous exercise provides some examples of self-injective rings.

(a) Show that R is self-injective if and only if every projective module is injective.

(b) Show that R is self-injective if and only if $M_n(R)$ is self-injective. (Hint: Use exercises at the end of (VIII).) Formulate this in context of Exercise (XII.5).

(c) Show that R is self-injective if and only if the basic ring of R is self-injective. (Hint: Use exercises at the end of (X).)

(d) If $R = \oplus \sum_{i=1}^{n} R_i$ then show R is self-injective if and only if each R_i is self-injective.

(e) Show that a semi-simple ring is self-injective.

(f) Determine the commutative self-injective rings.

(XII.12) EXERCISE. (Imbedding an R-module in an injective R-module.) Let M be an R-module. Let the characteristic of R be p^n. Thus R contains a copy of Z/Zp^n. For simplicity denote Z/Zp^n by S.

(a) Show that $\mathrm{Hom}_S(R,S)$ is an injective R-module. (Hint: $\mathrm{Hom}_S(R,S)$ is an R-module by defining $r\sigma$ by $(r\sigma)(r') = r(\sigma(r'$ Show $\mathrm{Hom}_R(N,\mathrm{Hom}_S(R,S)) \simeq \mathrm{Hom}_S(R \otimes_R N,S) \simeq \mathrm{Hom}_S(N,S)$. To show that $\mathrm{Hom}_S(R,S)$ is an injective R-module, we must show that whenever $M \to N$ is an injective R-morphism then $\mathrm{Hom}_R(N,\mathrm{Hom}_S(R,S)) \to \mathrm{Hom}_R(M,\mathrm{Hom}_S(R,S))$ is injective. But, by the above isomorphism, the latter map becomes $\mathrm{Hom}_S(N,S) \to \mathrm{Hom}_S(M,S)$. But S is self-injective, so this last map is injective.)

(b) Show that if R is a T-algebra and D is an injective T-module then $\mathrm{Hom}_T(R,D)$ is an injective R-module. (Hint: Use preceding proof.)

(c) Show M may be imbedded in an injective R-module. (Hint: Regard M as an S-module. Thus, $M \simeq \oplus \sum_{i=1}^{m} Z/Zp^{t_i}$ where $1 \le t_1 \le \cdots \le t_m \le n$. There is an S-injective morphism $\beta : M \to \oplus \sum_{i=1}^{m} S$ and $\oplus \sum_{i=1}^{m} S$ is an injective S-module. Now consider

$$M \simeq \mathrm{Hom}_R(R,M) \to \mathrm{Hom}_S(R,M) \to \mathrm{Hom}_S(R,\oplus \sum S)$$

where $m \to \sigma_m$ $(\sigma_m(1) = m)$, then σ_m (an R-morphism) is considered as an S-morphism and finally $\sigma_m \to \beta\sigma_m$. Check that the composition of these maps is an injective R-morphism.)

(XII.13) EXERCISE. (Characterization of injective R-modules with respect to the characteristic of R .) Let M be an injective R-module, p^n be the characteristic of R and let $S = Z/Zp^n$.

(a) Let e be a minimal idempotent of R . If $1 \le t \le n$, show $\mathrm{Hom}_S(eR,Z/Zp^t)$ is an indecomposable left R-module under $r\sigma$ given by $(r\sigma)(er') = \sigma(err')$.

(b) Let $Q = \mathrm{Hom}_S(R,M)$. Define $M \to Q$ by $m \to \phi_m$ where $\phi_m(r) = rm$. Show Q is naturally an R-module and that

$M \to Q$ is an injective R-morphism. Since M is injective deduce that M may be considered as an R-direct summand of Q.

(c) Considering M as an S-module, $M = \oplus \sum_{i=1}^{m} Z/Zp^{n_i}$ where $1 \le n_1 \le \cdots \le n_m \le n$. Show that

$$Q \simeq \oplus \sum_{i=1}^{m} \mathrm{Hom}_S(R, Z/Zp^{n_i}) .$$

(d) On the other hand, $R_R = \oplus \sum e_j R$ as a right-R-module where the e_j are minimal idempotents. Show that

$$Q \simeq \oplus \sum_{i,j} \mathrm{Hom}_S(e_j R, Z/Zp^{n_i})$$

where, by (a), $\mathrm{Hom}_S(e_j R, Z/Zp^{n_i})$ is indecomposable.

(e) Since M, by (b), is a direct summand of Q, apply the Krull-Schmidt Theorem and deduce that M is a direct sum of indecomposable modules of the form $\mathrm{Hom}_{Z/Zp^n}(eR, Z/Zp^t)$, $1 \le t \le n$.

<u>(XII.14)</u> <u>EXERCISE</u>. (Direct Sums of Cyclic Modules). (Warfield, [48]).

Let R be a ring.

(a) If $0 \to M \to N \to P \to 0$ is an exact sequence of R-modules, show the following are equivalent.

(1) For any r in R, the natural morphism $\mathrm{Hom}(R/Rr, N) \to \mathrm{Hom}(R/Rr, P)$ is surjective.

(2) For any r in R, $rM = M \cap rN$.

If (1) or (2) holds, M is called <u>relatively divisible</u> in N or an <u>RD-submodule</u> of M. A module Q is <u>RD-projective</u>

if whenever

$$\begin{array}{ccccc} & & Q & & \\ & & \downarrow \sigma & & \\ N & \xrightarrow{\beta} & P & \longrightarrow & 0 \end{array} \qquad \text{(row exact)}$$

with $M = \ker(\beta)$ an RD-submodule of N then there is a $\sigma^* : Q \to N$ with $\beta\sigma^* = \sigma$.

(b) Let P be any R-module. Show there is an exact sequence

$$0 \to M \to N \to P \to 0$$

where

(1) N is a direct sum of cyclic modules R/Rr for various r in R .

(2) M is an RD-submodule of N .

(Hint: Let S denote the set of cyclic R-modules. Let T denote the set of pairs $\langle Q,\sigma\rangle$ where $\sigma : Q \to P$ and Q is in S . Let $N = \oplus \sum Q$ be the direct sum over all such pairs and define $\Theta : N \to P$ by $\Theta|_Q = \sigma$; i.e., restriction of Θ to Q-th coordinate is σ . Since R is in S we have Θ surjective. Further if $M = \ker(\Theta)$ then M is an RD-submodule of N .)

(c) Show a R-module is RD-projective if and only if it is a direct summand of a direct sum of cyclic R-modules. (Hint: Use the above with P being the RD-projective module.)

(d) Let R be a commutative local ring. Show an R-module is RD-projective if and only if it is a direct sum of cyclic R-modules.

<u>(XII.15)</u> <u>EXERCISE</u>. (On indecomposable modules over local rings). Let R be a local ring with maximal ideal m of nilpotency β.

(a) If m is principal show the indecomposable R-modules are precisely $R/m^{\beta} = R$, $R/m^{\beta-1}$, ..., R/m.

(b) (Warfield). If the maximal ideal is not principal then for any $n > 0$ there exist indecomposable R-modules which cannot be generated by fewer than n elements. (For a hint one should see Warfield, [47], Theorem 2. The proof is similar in style to the proof of (XII.15).)

<u>(XII.16)</u> <u>EXERCISE</u>. (Menzin [35]). Let R be a ring and M a finitely generated R-module. By a <u>minimal presentation</u> of M we mean an exact sequence

$$(*) \qquad P_1 \xrightarrow{\sigma_1} P_0 \xrightarrow{\sigma_0} M \longrightarrow 0$$

where $P_0 \xrightarrow{\sigma_0} M$ and $P_1 \xrightarrow{\sigma_1} \ker(\sigma_0)$ are projective covers of M and $\ker(\sigma_0)$, respectively. Let M^* denote $\mathrm{Hom}_R(M,R)$. The sequence (*) induces naturally an exact sequence

$$0 \to M^* \to P_0^* \to P_1^* .$$

Define $T(M) = \mathrm{Coker}(P_0^* \to P_1^*) = P_1^*/\mathrm{Im}(P_0^*)$. For (*) show

(a) $T(M)$ is unique up to isomorphism and has no non-trivial projective direct summands.

(b) If M has no non-trivial projective direct summands then

$$P_0^* \to P_1^* \to T(M) \to 0$$

is a minimal presentation of T and $M \simeq T(T(M))$.

(c) If M is projective then $T(M) = 0$.

(d) $T(M_1 \oplus M_2) = T(M_1) \oplus T(M_2)$.

(e) If M has no non-trivial projective direct summands then M is indecomposable if and only if $T(M)$ is indecomposable.

The module $T(M)$ is called the transpose of M . The transpose has proven useful in generating large faithful indecomposable modules (see [35]).

(XII.17) EXERCISE. (Morita and Tachikawa [39]). Let R be a ring and F a free R-module of dimension n . Let M_1 and M_2 be two R-submodules of F and suppose there is an isomorphism $\phi : F/M_1 \to F/M_2$. Then there exists an isomorphism $\Phi : F \to F$ which induces the isomorphism ϕ and, in particular, M_1 and M_2 are isomorphic. (Hint: First prove this when R is semi-simple. See [39].)

(XII.18) EXERCISE. Let R be a ring with radical $N = \text{Rad}(R)$. Let M be a finitely generated R-module. By the semi-simple part of M we mean

$$S(M) = \{m \in M \mid Nm = 0\} .$$

(a) Let Q_1 and Q_2 be injective R-modules. Show that every diagram

$$\begin{array}{ccccc} 0 & \longrightarrow & M_1 & \xrightarrow{\beta_1} & Q_1 \qquad \text{(exact)} \\ & & \downarrow{\scriptstyle \phi} & & \\ 0 & \longrightarrow & M_2 & \xrightarrow{\beta_2} & Q_2 \qquad \text{(exact)} \end{array}$$

can be imbedded in a commutative diagram

$$\begin{array}{ccccc} 0 & \longrightarrow & M_1 & \xrightarrow{\beta_1} & Q_1 \\ & & \downarrow{\scriptstyle \phi} & & \downarrow{\scriptstyle \sigma} \\ 0 & \longrightarrow & M_2 & \xrightarrow{\beta_2} & Q_2 \end{array} .$$

(b) In (a), show that if $\mathrm{Im}(\beta_i) \supseteq S(Q_i)$ for $i = 1,2$ then ϕ is injective (bijective) if ϕ is injective (bijective).

(c) Show an injective R-module Q is indecomposable if and only if $S(Q)$ is simple.

(See [38].)

(XII.19) EXERCISE. Let M be a finitely generated projective R-module. If $N = \mathrm{Rad}(R)$ and M/NM is a free R/N-module, show M is a free R-module. Deduce that projective modules over local rings are free. [Hint: Use Nakayama's Lemma.]

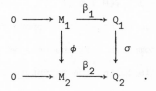

XIII. THE POLYNOMIAL RING R[X] PART I. REGULAR POLYNOMIALS

This is the first of several sections which deal with the theory of finite commutative rings. In particular our purpose is to develop the Galois theory of finite commutative local rings and show that this is as precise as that presented for finite fields in (II). The standard Galois theory for commutative rings describes a Galois correspondence between the group of ring automorphisms and its subgroups of a Galois extension S of a ring R and the separable subrings between S and R . In the finite setting a Galois extension S of R is a simple extension; that is, S is generated over R by a primitive element. This permits the use of the polynomial theory of R[X] in the description of automorphisms and ring extensions of R . It is the purpose of this section to begin a study of the polynomial ring R[X] and, in particular, regular polynomials (a subclass of which gives rise to the Galois extensions of

R), their factorizations and properties.

This section concerns lifting of polynomial theory from k[X] to R[X] and the structure of ideals in R[X] . Two short sections describe the R-algebra automorphisms of R[X] and the method of solution of a polynomial in R[X] extending the classical theory of higher order congruences.

Section (A). Regular Polynomials.

The decomposition of a commutative ring in (VI) induces a natural decomposition of the polynomial ring over a commutative ring. Thus, for most questions, the theory of polynomials over a commutative ring reduces to that of polynomials over a local commutative ring. Again, we remind the reader, the term "polynomial ring" is a violation of the convention that "ring" means "finite ring".

Throughout this section R denotes a finite local commutative ring with maximal ideal m and residue field $k = R/m$. We denote the natural projection $R[X] \to k[X]$ by μ . Thus the natural ring morphism $R \to R/m = k$ is simply the restriction of μ to the constant polynomials. We employ the usual terminology for polynomials; i.e., monic, degree, etc.

This section will survey and collect the basic theory of $R[X]$. This will concern the elements of $R[X]$ (units, primes, nilpotents, and, in particular, regular polynomials), the factorization in $R[X]$, and the lifting of the theory of $k[X]$ to $R[X]$. A rather complete treatment of polynomial algebras over a field is given by Greub's Linear Algebra (§XII).

(XIII.1) THEOREM. Let $R \subset S$ be commutative rings. Then if a is in S there exists a monic polynomial f in $R[X]$ such that $f(a) = 0$.

Proof. Let a be in S and consider $T = \{ \sum r_i a^i \mid r_i \text{ is in } R\}$. Then $R \subset T \subset S$ and for each element in T select a representative with least degree as a polynomial in a . Let B denote the set of these representatives and let m be the greatest degree of the polynomials in B . Since a^{m+1} is in T , $a^{m+1} = p(a)$ where $p(a)$ is in B . That is, a satisfies the monic polynomial $X^{m+1} - p(X)$.

The above result states that S is integral over R ; i.e., every a in S satisfies a monic polynomial in $R[X]$.

We let (f) denote the ideal generated by f .

If f and g are in $R[X]$, then

(a) f is nilpotent if there is an integer n with $f^n = 0$.

(b) f is a unit if there is a polynomial h with $fh = 1$.

(c) f is regular if f is not a zero divisor.

(d) f is <u>prime</u> if (f) is a proper prime ideal.

(e) f is <u>irreducible</u> if f is not a unit and whenever $f = gh$ then g or h is a unit.

(f) f is <u>primary</u> if (f) is a primary ideal.[1]

(g) f and g are <u>associated</u> if $(f) = (g)$.

(h) f and g are <u>coprime</u> if $R[X] = (f) + (g)$.

Utilizing the natural morphism $\mu : R[X] \to k[X]$, it is easy to show the following.

<u>(XIII.2)</u> <u>THEOREM</u>. Let $f = a_0 + a_1X + \cdots + a_nX^n$ be in $R[X]$.

(a) The following are equivalent:

(1) f is a unit,

(2) μf is a unit,

(3) a_0 is a unit and $a_1,\ldots,a_n$ are nilpotent.

(b) The following are equivalent:

(1) f is nilpotent,

(2) $\mu f = 0$,

(3) $a_0,\ldots,a_n$ are nilpotent,

(4) f is a zero divisor,

---

[1]An ideal I in R is <u>primary</u> if $I \neq R$ and whenever xy is in I then either x is in I or y^n is in I for some positive integer n .

(5) there is a non-zero a in R with $af = 0$.

(c) The following are equivalent:

(1) f is regular,

(2) $(a_0, a_1, \ldots, a_n) = R$,

(3) a_i is a unit for some i , $0 \leq i \leq n$,

(4) $\mu f \neq 0$.

If A is an ideal in R , we let $A[X]$ denote the set of polynomials in $R[X]$ of the form $a_0 + a_1 X + \cdots + a_n X^n$ where a_i is in A , $0 \leq i \leq n$. For the maximal ideal m of R we have the following characterizations of $m[X]$.

<u>(XIII.3)</u> <u>THEOREM</u>. (The Radical of $R[X]$).

(a) $m[X] = \bigcap \{P \mid P$ is a prime ideal in $R[X]\}$

(Nil or Prime Radical of $R[X]$)

(b) $m[X] = \{f \mid gf + 1$ is a unit for all g in $R[X]\}$

(Jacobson Radical of $R[X]$) .

We call $m[X]$ the <u>radical</u> of $R[X]$, denoted $\mathrm{Rad}(R[X])$.

<u>Proof</u>. For part (a) recall that $\bigcap \{P \mid P$ is a prime ideal in $R[X]\} = \{f$ in $R[X] \mid f$ is nilpotent$\}$. Then use (XIII.2) (b). For part (b), since the sum of a unit and a nilpotent is a unit, certainly $m[X] \subseteq \{f \mid gf + 1$ is a unit for all g in $R[X]\} = J$. Conversely, suppose f is in J where $f = a_0 + a_1 X + \cdots + a_n X^n$. Then $Xf + 1$ is a unit and thus, by (XIII.2) (a), $a_0, \ldots, a_n$ are nilpotent and

consequently in m .

Let f and g be in $R[X]$. The polynomial f is called a <u>divisor</u> of g if $(g) \subset (f)$ and a <u>proper divisor</u> if $(g) \subset (f)$ and $(g) \neq (f)$. It is easy to see that if g is regular then f is a proper divisor of g if and only if f is a divisor of g and μf is a proper divisor of μg .

<u>(XIII.4)</u> <u>THEOREM</u>. (Hensel's Lemma). Let f be in $R[X]$ and

$$\mu f = \bar{g}_1 \cdots \bar{g}_n$$

where $\bar{g}_1,\ldots,\bar{g}_n$ are pair-wise coprime. Then there exist $g_1,\ldots,g_n$ in $R[X]$ such that

(a) $g_1,\ldots,g_n$ are pair-wise coprime;

(b) $\mu g_i = \bar{g}_i$, $1 \le i \le n$;

(c) $f = g_1 \cdots g_n$.

<u>Proof</u>. We handle the case $n = 2$ then proceed by induction. For $n = 2$, $f = h_1h_2 + v$ where $\mu h_1 = \bar{g}_1$, $\mu h_2 = \bar{g}_2$ and v is in $m[X]$. Since $\bar{g}_1$ and $\bar{g}_2$ are coprime if and only if h_1 and h_2 are coprime, there exist λ_1 and λ_2 in $R[X]$ with $\lambda_1h_1 + \lambda_2h_2 = 1$. Take

$$h_{11} = h_1 + \lambda_2 v$$
$$h_{21} = h_2 + \lambda_1 v$$

and note

$$\begin{aligned} h_{11}h_{21} &= h_1h_2 + v(\lambda_1h_1 + \lambda_2h_2) + \lambda_1\lambda_2v^2 \\ &= h_1h_2 + v + \lambda_1\lambda_2v^2 \\ &= f + \lambda_1\lambda_2v^2 . \end{aligned}$$

Thus $f \equiv h_{11}h_{21} \bmod(v^2)$ where $\mu h_{i1} = \mu h_i$ for $i = 1,2$. It is clear that h_{11} and h_{21} are coprime and hence we may repeat the argument. Thus, for a positive integer t , there exist h_{1t} and h_{2t} such that

$$f \equiv h_{1t}h_{2t} \bmod(v^{2t})$$

and

$$\mu h_{it} = \mu h_i , \quad i = 1,2 .$$

But v is in $m[X]$ and thus nilpotent. Hence we may select t so that $f = h_{1t}h_{2t}$ finishing the case $n = 2$. The result follows by induction and the fact that if h_1 is coprime to h_i , $2 \le i \le n$, then h_1 and $h_2 \cdots h_n$ are coprime.

Before completing a discussion of factorization in $R[X]$ it is necessary to examine in greater detail prime, primary and regular elements in $R[X]$ and $k[X]$. We now show that each regular polynomial has a monic "representative".

<u>(XIII.5)</u> <u>LEMMA</u>. Let f be a regular polynomial in $R[X]$. Then there exists a sequence $\{f_j\}$ of monic polynomials in $R[X]$ with

$$\deg(f_j) = \deg(\mu f) ,$$

$$f_j \equiv f_{j+1} \pmod{m^j} ,$$

and for some g_j in $m[X]$ and unit b_j in R

$$b_j f \equiv f_j + g_j f_j \pmod{m^j} .$$

<u>Proof</u>. Let $f = \sum_{i=0}^{n} b_i X^i$ where $b_n \neq 0$ and $\deg(\mu f) = t \leq n$. Then b_t is a unit. Choose $g_1 = 0$ and $f_1 = b_t^{-1} f$. We now proceed by induction.

Assume $\{f_i\}_{i=1}^{j}$ have been selected to satisfy the lemma. Then $b_j f = f_j + g_j f_j + h$ where h is in $(m^j)[X]$. Since f_j is monic we may select q and r in $R[X]$ with $h = qf_j + r$ where $\deg(r) < \deg(f_j) = \deg(\mu f)$ or $r = 0$. Set $f_{j+1} = f_j + r$ and $g_{j+1} = g_j + q$.

We claim that g_{j+1} is in $m[X]$ and r is in $(m^j)[X]$. If $r = 0$ this statement is trivial. Otherwise suppose $f_j = a_0 + a_1 X + \cdots + a_{t-1} X^{t-1}$ and $q = c_0 + c_1 X + \cdots + c_s X^s$. In the product $f_j q$, the coefficient of X^{t+s} is c_s, of X^{t+s-1} is $c_s a_{t-1} + c_{s-1}$, etc. Since $h \equiv 0$ and $\deg(r) < \deg(f_j) = t$, it is easy to see that c_s, then c_{s-1}, then c_{s-2}, etc., are in m^j and consequently q is in $(m^j)[X]$. Then $r = h - qf_j$ is in $(m^j)[X]$.

Then

$$\begin{aligned} b_j f &= f_j + g_j f_j + h \\ &= (f_j + r) + (g_j + q)(f_j + r) - rg_j - rq \\ &= f_{j+1} + g_{j+1} f_{j+1} - r(g_j + q) \\ &\equiv f_{j+1} + g_{j+1} f_{j+1} \pmod{m^{j+1}} . \end{aligned}$$

<u>(XIII.6)</u> <u>THEOREM</u>. Let f be a regular polynomial in $R[X]$. Then there is a monic polynomial f^* with $\mu f = \mu f^*$ and, for an element a in R, $f(a) = 0$ if and only if $f^*(a) = 0$. Further, there is a unit v in $R[X]$ with $vf = f^*$.

<u>Proof</u>. Let m have nilpotency β. By the lemma $b_\beta f = f_\beta + g_\beta f_\beta = (1 + g_\beta)f$ where f_β is monic, b_β and $1 + g_\beta$ (since g_β is in $m[X]$) are units and $\mu f = \mu f_\beta$. Thus let $f^* = f_\beta$.

It is important to note several facts in the above proofs. First b_j in (XIII.5) and b_β in (XIII.6) do not depend on the index. If $f = \sum_{i=0}^{n} a_i X^i$ and $\deg(\mu f) = t \le n$, then b_β may be taken to be a_t. In the proof of (XIII.6) $b_\beta f = (1 + g_\beta) f_\beta$ where b_β and $1 + g_\beta$ are units, hence $(f) = (f_\beta)$. Finally, f_β is obtained in at most β steps. Thus most questions on regular polynomials may be reduced to questions concerning monic polynomials.

Let J denote the set of all polynomials f in $R[X]$ such that μf has distinct zeros in the algebraic closure of k. The class J has been useful in certain problems in combinatorial mathematics.

We propose now to classify the regular irreducible polynomials in R[X] .

Suppose that f is regular in R[X] and μf is irreducible (equivalently, prime) in k[X] . Then, if $f = gh$ either μg or μh is a unit. By (XIII.2) g or h is a unit. Thus, if μf is irreducible then f is irreducible.

On the other hand, if f is irreducible in R[X] suppose $\mu f = \delta g_1^{e_1} \cdots g_t^{e_t}$ where δ is a unit and the g_i are monic irreducible coprime polynomials in k[X] . Then, unless $t = 1$, by Hensel's Lemma f factors non-trivially. That is, if f is irreducible then $\mu f = \delta g^n$ where δ is a unit in k and g is irreducible.

<u>(XIII.7)</u> <u>THEOREM</u>. (On Irreducibles). Let f be a regular polynomial in R[X] . Then

(a) If μf is irreducible in k[X] then f is irreducible.

(b) If f is irreducible then $\mu f = \delta g^n$ where δ is in k and g is a monic irreducible in k[X] .

(c) If f is in J then f is irreducible if and only if μf is irreducible.

Prime polynomials in R[X] are irreducible; however, an irreducible polynomial need not be prime. We now show that this property distinguishes finite fields and finite local rings which are not fields. This provides an interesting characterization of finite fields.

(XIII.8) LEMMA. Let f be a regular irreducible polynomial in J. Then f is prime if and only if $m \subseteq (f)$.

Proof. If f is prime then $R[X]/(f)$ is a finite field. But if a is in m then the coset $a + (f)$ is nilpotent in $R[X]/(f)$. Thus a is in (f).

Conversely, if $m \subseteq (f)$ then $m[X] \subseteq (f)$. Suppose that the coset $g + (f)$ is nilpotent in $R[X]/(f)$. Then f divides g^n for some n and thus μf divides $(\mu g)^n$. Since f is irreducible and in J we have that μf divides μg; i.e., $\mu g = \bar{h}(\mu f)$. Let h be a pre-image in $R[X]$ of $\bar{h}$. Then $hf = g + j$ where j is in $m[X]$. Hence g is in (f) and, consequently, $R[X]/(f)$ is a field.

(XIII.9) THEOREM. (A Characterization of Finite Fields). Let R be a finite local commutative ring. The following are equivalent:

(a) R is a finite field.

(b) Every regular irreducible polynomial in $R[X]$ is prime.

(c) There exists at least one regular irreducible polynomial in J which is prime.

Proof. Clearly (a) implies (b) and (b) implies (c). It remains to show (c) implies (a). Suppose R is not a field and f is a regular irreducible polynomial in J. Let $a \neq 0$ be in m. If a is in (f) then f divides a which is impossible by observing their degrees and that f is regular. Thus m is not in (f) and by (XIII.8) f is not prime.

(XIII.10) COROLLARY. Let R be a local ring which is not a field. Then

(a) $R[X]$ has no principal prime ideals generated by regular polynomials.

(b) $R[X]$ contains no regular irreducible polynomials which are prime.

If A and B are ideals in $R[X]$ then $A + B$ is the ideal generated by A and B. If $B = (f)$ then $A + B$ is denoted by (A,f). If $N \subseteq m$ then $N[X]$ is denoted by (N) in $R[X]$.

An alternate approach to (XIII.9) and (XIII.10) and interesting in its own right is provided by the following observations.

If N is a maximal ideal in $R[X]$ then $N \cap R = m$. The image of N under $\mu : R[X] \to k[X]$ is $(\bar{f})$ where $\bar{f}$ is an irreducible polynomial in $k[X]$. From this it is easy to deduce that every maximal ideal of $R[X]$ is of the form (m,f) where μf is an irreducible polynomial in $k[X]$; i.e., the maximal ideal is generated by m and f in $R[X]$.

Consequently, f is an irreducible polynomial in J. Thus, for an irreducible polynomial f in J, the ideal (f) is maximal if and only if $m = (0)$; i.e., R is a finite field.

Let S denote either R or $R[X]$. An ideal A in S is called _primary_ if A is a proper ideal and whenever ab is in A then either a is in A or b^n is in A for some integer n. The _radical_ of an ideal A of S, denoted $r(A)$, consists of all elements a of S such that a^n is in A for some integer n. The image of an ideal A under $R[X] \to k[X]$ is denoted by μA. Finally, a proper ideal A is called _regular_ if $\mu A \neq (0)$. Note that if A is not regular then A is nilpotent, and conversely.

We return now to factorization of the regular elements of $R[X]$ and complete the development begun in (XIII.4).

(XIII.11) _THEOREM_. (Factorization Theorem). Let f be a regular polynomial in $R[X]$. Then

(a) $f = \delta g_1 \cdots g_n$ where δ is a unit and $g_1, \ldots, g_n$ are regular primary coprime polynomials.

(b) If $f = \delta g_1 \cdots g_n = \beta h_1 \cdots h_m$ where δ and β are units and $\{g_i\}$ and $\{h_j\}$ are regular primary coprime polynomials then $n = m$ and, after renumbering, $(h_i) = (g_i)$, $1 \leq i \leq n$.

Proof. Let f be regular in $R[X]$. Then $\mu f \neq 0$ thus $\mu f = \bar{\delta}\,\bar{\Pi}_1^{h_1} \cdots \bar{\Pi}_n^{h_n}$ where $\bar{\delta}$ is a unit and $\bar{\Pi}_1,\ldots,\bar{\Pi}_n$ are irreducible coprime polynomials in $k[X]$. Then $\bar{\Pi}_1^{h_1},\ldots,\bar{\Pi}_n^{h_n}$ are regular coprime primary elements in $k[X]$. By (XIII.4) $f = \delta\,\Pi_1 \cdots \Pi_n$ where $\mu\delta = \bar{\delta}$ and $\mu\Pi_i = \bar{\Pi}_i^{h_i}$. From Exercise (XIII.3) it is easy to see that the Π_i are regular primary coprime polynomials.

The second assertion is standard. That is, in terms of principal ideals, if

$$(g_1) \cdots (g_n) = (h_1) \cdots (h_m)$$

where the $\{(g_i)\}$ and $\{(h_j)\}$ are regular primary coprime ideals then $n = m$ and after a suitable ordering $(g_i) = (h_i)$, $1 \le i \le n$.

To complete this discussion of factorization we examine the regular primary polynomials in greater detail. A regular polynomial f is primary if and only if μf is primary in $k[X]$. Equivalently, $\mu f = \bar{\delta}\,\bar{\Pi}^h$ where $\bar{\delta}$ is a unit in k and $\bar{\Pi}$ is an irreducible polynomial in $k[X]$. We say that an irreducible polynomial Π in $R[X]$ is a basic irreducible if $\mu\Pi$ is irreducible in $k[X]$.

(XIII.12) PROPOSITION. An element f of $R[X]$ is a primary regular non-unit if and only if $f = \delta\,\Pi^h + \beta$ where δ is a unit, Π is a basic irreducible, $h \ge 1$ and β is in $m[X]$.

The uniqueness of the above expression for a primary regular non-unit is examined in the exercises in this section.

Following (XIII.10) were some comments on maximal ideals in $R[X]$. We now consider regular ideals from that viewpoint. Let A be a regular ideal in $R[X]$. Then $\mu A = (\bar{f})$ where $\bar{f}$ is a non-zero non-unit in $k[X]$. Set $N = A \cap (m[X])$. Then if $\mu f = \bar{f}$ we have that $A = (N,f)$.

Conversely if N is an ideal in $m[X]$ and f is a regular non-unit in $R[X]$ then (N,f) is a regular ideal. A regular ideal $A = (N,f)$ is primary if and only if (μf) is a primary ideal in $k[X]$; i.e., $\mu f = \bar{\delta}\, \bar{\Pi}^h$ where $\bar{\delta}$ is non-zero in k and $\bar{\Pi}$ is an irreducible polynomial in $k[X]$. In this case, by (XIII.12), $f = \delta\, \Pi^h + \beta$ and $r(A) = (m,\Pi)$. Finally, since an ideal is prime if and only if the ideal is primary and equal to its radical we have the following result.

<u>(XIII.13)</u> <u>THEOREM</u>.

(a) An ideal Q of $R[X]$ is a regular primary ideal if and only if $Q = (N,\delta\, \Pi^h + \beta)$ where N is a nilpotent ideal in $m[X]$, δ is a unit, Π is a basic irreducible, $h \geq 1$ and β is in $m[X]$. In this case, $r(Q) = (m,\Pi)$.

(b) An ideal P of $R[X]$ is prime if and only if $P = (m,\Pi)$ where Π is a basic irreducible or $\Pi = 0$. If Π is a basic irreducible then P is a maximal ideal and this gives all maximal ideals.

Proof. For part (b) one needs to observe that the only non-regular prime ideal in $R[X]$ is $m[X]$.

There are numerous nilpotent primary ideals. If N is any ideal in m , then $N[X]$ is a nilpotent primary ideal with radical $m[X]$.

Section (B). R-Algebra Automorphisms of $R[X]$.

In this section we determine the R-algebra automorphisms $R[X] \to R[X]$ where R is a commutative local ring with maximal ideal m and residue field $k = R/m$. These results were developed for arbitrary commutative rings by Gilmer [66]. Coleman and Enochs [62] have shown them to be important in questions of isomorphic polynomial rings.

(XIII.14) LEMMA. Let f be a non-zero non-unit in $k[X]$ of degree n . Let g be a polynomial in $k[X]$ of degree m . Then $g(f)$ is a polynomial of degree mn . Further f generates $k[X]$ over k if and only if $\deg(f) = 1$; i.e., $f = a_0 + a_1X$ where $a_1 \neq 0$.

Proof. The first statement is straightforward. If f is a generator of $k[X]$ over k then $X = g(f)$ for some g in $k[X]$. Then, if $\deg(f) = n$ and $\deg(g) = m$, $mn = 1$ and thus $f(X) = a_0 + a_1X$, $a_1 \neq 0$. Conversely, if $f(X) = a_0 + a_1X$ then X is in $k[f]$. Thus $k[X] = k[f]$.

We have immediately.

(XIII.15) THEOREM. Every k-algebra automorphism $\sigma : k[X] \to k[X]$ is of the form

$$\sigma : X \to a_0 + a_1X$$

where $a_1 \neq 0$.

Consider now an R-algebra morphism $\sigma : R[X] \to R[X]$. It is clear that the action of σ is uniquely determined by the image of X under σ . Thus if f is in $R[X]$ the R-algebra morphism $R[X] \to R[X]$ induced by $X \to f$ is denoted by σ_f .

If σ_f is an R-automorphism then units map to units and nilpotents map to nilpotents under σ_f . In particular, the ideal $m[X]$ is characteristic; i.e., $\sigma_f(m[X]) \subseteq m[X]$. This implies that σ_f induces a k-algebra automorphism $\bar{\sigma}_f : k[X] \to k[X]$ by $\bar{\sigma}_f(\bar{h}) = \mu(\sigma_f(h))$ where h is a pre-image of $\bar{h}$. It is clear then that if $f = a_0 + a_1X + \cdots + a_nX^n$ in $R[X]$ then $\bar{\sigma}_f = \sigma_{\mu f}$; i.e., σ_f induces the k-automorphism

$$\sigma_{\mu f} : X \to \mu f .$$

By (XIII.15) $\mu f = (\mu a_0) + (\mu a_1)X$ where $\mu a_1 \neq 0$. We conclude that a_1 is a unit and $a_2,\ldots,a_n$ are nilpotent.

<u>(XIII.16)</u> <u>THEOREM</u>. Let $f = a_0 + a_1X + \cdots + a_nX^n$ be in $R[X]$. Then $\sigma_f : X \to f$ induces an R-algebra automorphism of $R[X]$ if and only if a_1 is a unit and $a_2,\ldots,a_n$ are nilpotent. Each R-algebra automorphism of $R[X]$ is of the form σ_f for some such f .

<u>Proof</u>. It is only necessary to show that if $f = a_0 + a_1X + \cdots + a_nX^n$ where a_1 is a unit and $a_2,\ldots,a_n$ are nilpotent then σ_f is an R-algebra automorphism of $R[X]$. We first show that X is in $R[f]$ and thus σ_f is surjective. Since $R[f] = R[a_1^{-1}(f - a_0)]$ we may without loss suppose that $a_0 = 0$ and $a_1 = 1$. Thus $f = X + a_2X^2 + \cdots + a_nX^n$ where $a_2,\ldots,a_n$ are in m . Form

$$g = f - a_2f^2 - \cdots - a_nf^n .$$

Then it is easy to see that $g = X + b_2X^2 + \cdots + b_mX^m$ where b_i is in m^2 for $2 \leq i \leq m$. Since m is nilpotent by repeating the process we obtain X .

To conclude the proof we show σ_f is injective. It is easy to see that σ_f is injective if and only if σ_{f-a_0} is injective thus we may assume $a_0 = 0$. Suppose $g = g_0 + g_1X + \cdots + g_mX^m$ is in $R[X]$ and $\sigma_f(g) = 0$. Since $a_0 = 0$ we must have $g_0 = 0$. Then

$$(g_1 + g_2f + \cdots + g_mf^{m-1})f = 0 .$$

But, since a_1 is a unit, f is not a zero divisor. Thus $g_1 + g_2f + \cdots + g_mf^{m-1} = 0$. Repeating the argument shows that

$g_1 = 0$, then $g_2 = 0$, etc.; that is, $g = 0$ and σ_f is injective.

To conclude this section, suppose $\phi : R \to R$ is a ring automorphism and f is a polynomial in $R[X]$. Define

$$\sigma_{\phi,f} : R[X] \to R[X]$$

by

$$(\sigma_{\phi,f})(\sum a_i X^i) = \sum \phi(a_i) f^i .$$

<u>(XIII.17)</u> <u>THEOREM</u>.

(a) $\sigma_{\phi,f}$ is injective if and only if σ_f is injective.

(b) $\sigma_{\phi,f}$ is surjective if and only if σ_f is surjective.

<u>Proof</u>. Note if σ_f is injective then $\sigma_{\phi,f}$ is injective. Conversely, if $\sigma_{\phi,f}$ is injective and $\sum a_i f^i = 0$ then $\sigma_{\phi,f}(\sum \phi^{-1}(a_i)X^i) = 0$. Thus $\phi^{-1}(a_i) = 0$, and hence $a_i = 0$ for each i . Since the image of $\sigma_{\phi,f}$ is $R[f]$ the second statement is easy.

<u>Section (C)</u>. <u>Solutions of a Polynomial</u>.

To analyze the method of solution of a polynomial in $R[X]$ we extend the classical theory of higher order congruences. Again, R denotes a commutative local ring with maximal ideal m and residue field $k = R/m$.

Let β denote the nilpotency of m ; i.e., β is the least positive integer with $m^\beta = 0$. There is a natural sequence of ring morphisms

$$R = R/m^\beta \xrightarrow{\sigma_\beta} R/m^{\beta-1} \longrightarrow \cdots \longrightarrow R/m \xrightarrow{\sigma_1} 0$$

where $\sigma_i : R/m^i \to R/m^{i-1}$. For each i there is also a natural morphism $\mu_i : R/m^i \to k$. The kernel of σ_i is a k-vector space m^{i-1}/m^i where the action of k on m^{i-1}/m^i is given by $\bar{\alpha}n = \alpha n$ where n is in m^{i-1}/m^i and $\mu_i(\alpha) = \bar{\alpha}$.

For simplicity we suppress the subscripts denoting σ_i and μ_i by σ and μ , respectively. Likewise we denote the extensions of σ_i and μ_i to the polynomial rings by σ and μ .

The purpose is to construct the solutions of f in $(R/m^i)[X]$ from the solutions of σf in $(R/m^{i-1})[X]$. Let $\dim_k(m^{i-1}/m^i) = t$ and $\{v_1,\ldots,v_t\}$ be a fixed k-basis for m^{i-1}/m^i . We now construct solutions of f in $(R/m^i)[X]$ from solutions of σf in $(R/m^{i-1})[X]$.

Let $\bar{a}$ in R/m^{i-1} be a solution of σf and suppose $\sigma a = \bar{a}$. Let $b = a + \eta$ for some η in m^{i-1}/m^i . The object is to select η so that $f(b) = 0$ in R/m^i .

Since $(m^{i-1}/m^i)^2 = 0$,

$$\begin{aligned} f(b) &= f(a + \eta) \\ &= f(a) + \eta f'(a) + \eta^2 Q \\ &= f(a) + \eta f'(a) \end{aligned}$$

where f' is the formal derivative of f and Q is in R/m^i.

Requiring that $f(b) = 0$ implies

$$f(a) = -\eta f'(a) = -\mu(f'(a))\eta$$

since η is in m^{i-1}/m^i. Further, since $(\sigma f)(\bar{a}) = 0$, $f(a)$ is in m^{i-1}/m^i. Thus relative to the k-basis $\{v_1,\ldots,v_t\}$,

$$f(a) = \sum b_i v_i ,$$
$$-\eta = \sum a_i v_i$$

where a_i and b_i are in k. Hence

$$0 = \sum b_i v_i + \mu(f'(a))(\sum a_i v_i)$$
$$= \sum (b_i + \mu(f'(a))a_i)v_i .$$

For each i we must have $0 = b_i + \mu(f'(a))a_i$.

Three cases arise:

(a) $f'(a)$ is a unit. In this case, $\mu(f'(a)) \neq 0$ and each a_i is uniquely determined. Thus there is exactly one solution a in R/m^i of f with $\sigma a = \bar{a}$.

(b) $f'(a)$ is in m/m^i and there exists a $b_j \neq 0$. In this case there are no solutions a of f with $\sigma a = \bar{a}$.

(c) $f'(a)$ is in m/m^i and $b_j = 0$ for all j. In this case $f(a) = 0$ for any a with $\sigma a = \bar{a}$. Thus, there exist

$$|m^{i-1}/m^i| = |k|^t$$

solutions a of f with $\sigma a = \bar{a}$.

Observe we obtain all solutions f in this manner since if $f(a) = 0$ for f in $(R/m^i)[X]$ and a in R/m^i then σa satisfies σf in $(R/m^{i-1})[X]$. This reduces in general the question of solutions of f in R to solutions of μf in k .

Many of the following exercises together with additonal results may be found in Snapper [57], [58], [59] and [60].

<u>(XIII.1)</u> <u>EXERCISE</u>. Let A and B be ideals in $R[X]$ with $A + B = R[X]$ and $A = (a_1,\ldots,a_n)$. Show by induction that $(a_1^t,\ldots,a_n^t) + B = R[X]$ for $t = 1,2,3,\ldots$. In particular, if f and g are coprime then f^t and g^s are coprime for positive integers t and s .

<u>(XIII.2)</u> <u>EXERCISE</u>. Prove Theorem (XIII.2).

<u>(XIII.3)</u> <u>EXERCISE</u>. Let A and B be ideals in $R[X]$. Show:

(a) A is prime if and only if $m[X] \subset A$ and μA is prime.

(b) A is maximal if and only if $m[X] \subset A$ and μA is maximal.

(c) A regular ideal is prime if and only if it is maximal.

(d) (1) $r(\mu A) = \mu(r(A))$.

(2) $r(A) \subset r(B)$ if and only if $r(\mu A) \subset r(\mu B)$.

(3) $r(\mu A) = r(\mu B)$ if and only if $r(A) = r(B)$.

(4) If $\mu A = \mu B$ then $r(A) = r(B)$.

(e) A regular ideal is primary if and only if $r(A)$ is prime.

(XIII.4) EXERCISE. Let f be a regular primary polynomial in $R[X]$. If $f = \delta_1 \Pi_1^{h_1} + \beta_1 = \delta_2 \Pi_2^{h_2} + \beta_2$ (XIII.12) show $h_1 = h_2$ and $(\Pi_1) = (\Pi_2)$.

(XIII.5) EXERCISE. Show that if f and g are regular associated elements in $R[X]$ then $f = \delta g$ where δ is a unit.

(XIII.6) EXERCISE. (Euclidean Algorithm). Let f and g be non-zero polynomials in $R[X]$. If g is regular show there exist polynomials q and r in $R[X]$ with

$$f = gq + r$$

and $\deg(r) < \deg(g)$. [Hint: Replace g by a monic associative.] Are q and r unique?

(XIII.7) EXERCISE. Let $R \subset S$ and a be in S . By (XIII.1) there is a monic polynomial f in $R[X]$ with $f(a) = 0$. Show there is a natural surjective R-algebra morphism $R[X]/(f) \to R[a]$.

(XIII.8) EXERCISE. Let f be a regular element in $R[X]$. Then show that g is a proper divisor of f if and only if g is a divisor of

f and μg is a proper divisor of μf .

(XIII.9) EXERCISE. (Residue Class Rings). Let A be an ideal of $R[X]$ and let $\sigma : R[X] \to R[X]/A$ denote the natural ring morphism. Let $\sigma(r(A)) = N$. Show that

$$\begin{aligned} N &= r(\sigma(0)) \\ &= r(A)/A \quad \text{and} \end{aligned}$$

$$(R[X]/A)/N \simeq (R[X]/A)/(r(A)/A) \simeq$$

$$(R[X])/r(A) \simeq (R[X]/m[X])/(r(A)/m[X]) \ .$$

In particular,

$$R[X]/(m,f) \simeq k[X]/(\mu f)$$

and if $Q = (N, \delta\, \Pi^h + \beta)$ is regular primary ideal, then $R[X]/Q$ is a finite local ring with residue field $k[X]/(\mu\Pi)$.

(XIII.10) EXERCISE. Show that if f and g are monic regular polynomials and $(f) = (g)$; i.e., f and g are associates then $f = g$. Restate (XIII.11) if the primary polynomials are chosen to be monic.

(XIII.11) EXERCISE. Show every ideal in $R[X]$ is finitely generated.

(XIII.12) EXERCISE. Let A be an ideal in $R[X]$. Show that $R[X]/A$ is a finitely generated free R-module if and only if $A = (f)$ where f is a monic polynomial.

(XIII.13) EXERCISE. The order of a regular polynomial f, denoted $O(f)$, is the minimum degree of non-zero polynomials in (f).

(a) If f is a regular polynomial in $R[X]$, show there exists a non-zero r in R with $O(f) = \deg(rf)$. (Hint: Note (XIII.6) and the comment after.)

(b) If $f = r_n X^n + \cdots + r_m X^m + \cdots + r_1 X + r_0$ is a regular polynomial, show $O(f)$ is the exponent m of lowest power in X in f such that $f - (r_m X^m + \cdots + r_0)$ is a zero divisor.

(XIII.14) EXERCISE. Let A be an ideal in $R[X]$. If the residue class ring $R[X]/A$ is finitely generated over $R/R \cap A$ then A is said to have finite degree.

(a) Show A has finite degree if and only if A contains a monic polynomial.

(b) Show A has finite degree if and only if μA has finite degree in $k[X]$.

(c) Show A has finite degree if and only if A is regular, say $A = (N,f)$. In this case, show $R[X]/A$ has a minimal generating set over $R/R \cap A$ of $O(f)$ elements.

(XIII.15) EXERCISE. (Fields). Let $R[[X]]$ denote the ring of formal power series over R. If $f = \sum_{i=0}^{\infty} a_i X^i$, show that f is nilpotent if and only if each a_i is nilpotent.

(XIII.16) EXERCISE. (Glimer). For a polynomial f in $R[X]$ let A_f denote the ideal in R generated by the coefficients of f. Show for polynomials f and g and r in R:

(a) $A_{f+g} \subseteq A_f + A_g$

(b) $A_{fg} \subseteq A_f A_g$

(c) $A_{rf} = rA_f$.

(XIII.17) EXERCISE. (Notation of (B)). Let f be in $R[X]$, r be in R and u a unit of R . Show

(a) The following are equivalent:

(1) σ_f is surjective

(2) σ_{r+f} is surjective

(3) σ_{uf} is surjective.

(b) The following are equivalent:

(1) σ_f is injective

(2) σ_{r+f} is injective

(3) σ_{uf} is injective.

(XIII.18) EXERCISE. (Notation of (B)). Let $f = a_0 + a_1X + \cdots + a_nX^n$. Show that if σ_f is surjective then a_1 is a unit and $a_2,\ldots,a_n$ are nilpotent. That is, show that if σ_f is surjective then it is injective and thus an R-algebra automorphism of $R[X]$.

XIV. THE POLYNOMIAL RING R[X]

PART II. SEPARABLE LOCAL EXTENSIONS

In this section we complete the polynomial theory and extension theory necessary to describe the Galois extensions of local rings presented in the next section. Our purpose is to show that if $R \subseteq S$ are local commutative rings then the following are equivalent:

(a) S is projective as a module over $S^e = S \otimes_R S$.

(b) The exact sequence of natural S^e-modules $0 \to \ker(\phi) \to S^e \xrightarrow{\phi} S \to 0$ splits where $\phi(s_1 \otimes s_2) = s_1 s_2$.

(c) There is an idempotent e in S^e such that $\ker(\phi)e = 0$ and $\phi(e) = 1$.

(d) $\mathrm{Hom}_{S^e}(S, \ )$ is cokernel preserving.

(e) S is unramified over R.

(f) $S \simeq R[X]/(f)$ where f is a basic irreducible in $R[X]$.

We begin the development of a Galois theory for commutative local rings. The last section described much of the necessary polynomial theory. This section will examine local simple extensions of local commutative rings.

Let R be a commutative ring (not necessarily local). A ring S is called an R-algebra if there is a ring morphism $R \to$ center of S. The image of R is denoted by $R \cdot 1$. To an R-algebra S we associate a second algebra $S \otimes_R S^0$ called the enveloping algebra of S and denoted by S^e.[1] It is customary to examine S as a left module over S^e where $(a_1 \otimes a_2)a = a_1 a a_2$. Before specializing S to be a commutative extension of R we examine S and S^e.

There is a natural surjective S^e-morphism

$$\phi : S^e \to S$$

given by $\phi(a_1 \otimes a_2) = a_1 a_2$. This gives rise to an exact sequence of S^e-modules

$$0 \longrightarrow \ker(\phi) \xrightarrow{i} S^e \xrightarrow{\phi} S \longrightarrow 0 .$$

(Recall, $0 \longrightarrow M \xrightarrow{\alpha} N \xrightarrow{\beta} P \longrightarrow 0$ is an exact sequence of modules and morphisms if $\mathrm{Im}(\alpha) = \ker(\beta)$, α is injective and β is surjective.) The kernel of ϕ, $\ker(\phi)$, is the left ideal of S^e generated by

---

[1]Multiplication in $S \otimes_R S^0$ is given by $(s_1 \otimes t_1)(s_2 \otimes t_2) = (s_1 s_2) \otimes (t_1 t_2)$. The exercises in (XII) examine the tensor product $\otimes$.

$\{a \otimes 1 - 1 \otimes a \mid a \text{ in } S\}$. The algebra S is called R-separable if S is projective as a module over S^e .

(XIV.1) THEOREM. The following are equivalent.

(a) S is a separable R-algebra.

(b) $0 \longrightarrow \ker(\phi) \longrightarrow S^e \xrightarrow{\phi} S \longrightarrow 0$ splits; i.e., there is an S^e-morphism $\psi : S \to S^e$ such that $\phi\psi$ = identity on S .

(c) There is an element e in S^e such that $\ker(\phi)e = 0$ and $\phi(e) = 1$.

Proof. It is easy to show that (a) and (b) are equivalent. (Note: to show (b) implies (a) use the observation that if the sequence splits then S may be considered as an S^e-direct summand of the free S^e-module S^e .) Now assume (b); i.e., the sequence splits. Then there is an S^e-morphism $\psi : S \to S^e$ such that $\phi\psi$ = identity on S . Let $e = \psi(1)$. One checks that e is the desired element in (c).

Conversely, suppose e is in S^e such that $\ker(\phi)e = 0$ and $\phi(e) = 1$. Define $\psi : S \to S^e$ by $\psi : a \to (a \otimes 1)e = (1 \otimes a)e$. One checks that ψ is the desired S^e-morphism.

An element e in S^e satisfying (c) above is called a separability idempotent for S . Note that $\phi(e - 1 \otimes 1) = 0$ so $e^2 - e = (e - (1 \otimes 1))e$ is in $\ker(\phi)e$ and thus equal to 0 ; i.e., e is an idempotent.

To describe separability from a different viewpoint consider the correspondence

$$N \to \mathrm{Hom}_{S^e}(S,N)$$

where N is a left S^e-module. It is easy to see that $\mathrm{Hom}_{S^e}(S,N)$ is an R-module. Further, if $\sigma : N \to N'$ is an S^e-morphism then

$$\sigma^* : \mathrm{Hom}_{S^e}(S,N) \to \mathrm{Hom}_{S^e}(S,N')$$

where $\sigma^*(f) = \sigma f$ for f in $\mathrm{Hom}_{S^e}(S,N)$ is an R-morphism. The correspondence is called <u>cokernel preserving</u> or <u>right exact</u> if σ^* is surjective whenever σ is surjective. The above correspondence is denoted by $\mathrm{Hom}_{S^e}(S, \ )$.

It is simple to use the properties of projective modules to show:

<u>(XIV.2)</u> <u>THEOREM</u>. The R-algebra S is R-separable if and only if $\mathrm{Hom}_{S^e}(S, \ )$ is cokernel preserving.

o o o

<u>NOTE</u>: For the remainder of this section S and R denote commutative local rings with maximal ideals M and m and residue fields $K = S/M$ and $k = R/m$, respectively.

o o o

We say that S is an _extension_ of R if R is a subring of S . Observe that if S is an extension of R then S is an R-algebra with $R \cdot 1 = R$ and $S^e = S \otimes_R S$.

The local ring S is an _unramified_ extension of R if M = Sm ; i.e., m generates the maximal ideal of S . The first purpose of this section is to show that S is an unramified extension of R if and only if S is a separable extension of R . Several results are preliminary to this.

(XIV.3) _THEOREM_. The local ring S is a separable extension of R if and only if S/mS is a separable R/m-algebra.

Proof. The proof is involved. We first show that S/mS is a separable R-algebra when S is a separable extension of R . To do this we show by (XIV.2) that

$$\mathrm{Hom}_{(S/mS)^e}(S/mS,\ )$$

is cokernel preserving. Let N be an $(S/mS)^e$-module. Then N may be considered as an S^e-module by

$$(s_1 \otimes s_2)n = [(s_1 + mS) \otimes (s_2 + mS)n] .$$

The proof will be complete if we show

$$\mathrm{Hom}_{(S/mS)^e}(S/mS,N) \simeq \mathrm{Hom}_{S^e}(S,N) \qquad \text{(as groups)}.$$

Let f be in $\mathrm{Hom}_{S^e}(S,N)$. Then $f(1)$ is in N. Define

$$\psi : \mathrm{Hom}_{S^e}(S,N) \to \mathrm{Hom}_{(S/mS)^e}(S/mS,N)$$

by

$$\psi(f)(s + mS) = (s + mS) \otimes (1 + mS)f(1) .$$

One checks that $\psi(f)$ is in $\mathrm{Hom}_{(S/mS)^e}(S/mS,N)$.

If $g : S/mS \to N$ is an $(S/mS)^e$-morphism then define

$$\phi : \mathrm{Hom}_{(S/mS)^e}(S/mS,N) \to \mathrm{Hom}_{S^e}(S,N)$$

by $\phi(g)(s) = g(s + mS)$. It is easy to check that $\phi(g)$ is in $\mathrm{Hom}_{S^e}(S,N)$ and that $\phi\psi(f) = f$ and $\psi\phi(g) = g$. Thus S/mS is a separable R-algebra.

Suppose T is an R-algebra and I is an ideal of R with $I \cdot 1 = 0$. Then T is naturally an R/I-algebra. Further

$$T \otimes_R T^0 = T \otimes_{(R/I)} T^0 ,$$

so T is R-separable if and only if T is R/I-separable. We often use this "shift of coefficients."

We have shown that if S is R-separable then S/mS is R-separable. Consequently, by the preceding remark, S/mS is a separable R/m-algebra.

Assume now that S/mS is a separable R/m-algebra. Consider the following commutative diagram:

$$\begin{array}{ccccc} S^e & \xrightarrow{\phi} & S & \longrightarrow & 0 \\ \downarrow & & \downarrow & & \\ (S/mS)^e & \xrightarrow{\bar{\phi}} & (S/mS) & \longrightarrow & 0 \end{array} .$$

Since S/mS is R/m-separable, $(S/mS)^e$ contains a separability idempotent $\bar{e}$. This idempotent $\bar{e}$ may be lifted (as in (VII)) to an idempotent f in S^e. Since $\bar{\phi}(\bar{e}) = 1$, by the commutativity of the above diagram $\phi(f)$ is in $1 + mS$. But $mS \subseteq M$ (maximal ideal of S) and $1 + mS$ consists of units. Since $\phi(f)$ is a unit, set

$$e = (\phi(f)^{-1} \otimes 1)f(\phi(f) \otimes 1) .$$

Then e is an idempotent and $e \to \bar{e}$. Further, one checks that $\phi(e) = 1$.

As R-modules the sequence

$$0 \to \ker(\phi) \to S^e \xrightarrow{\phi} S \to 0$$

splits with $S \to S^e$ given by $s \to s \otimes 1$. Then $\Pi : t \to t - \phi(t) \otimes 1$ is a projection (as R-modules) of S^e onto $\ker(\phi)$.

The composition of

$$(S^e)e \xrightarrow{i} S^e \xrightarrow{\Pi} \ker(\phi) \xrightarrow{()e} \ker(\phi)e$$

given by

$$te \to (t - \phi(t) \otimes 1)e$$

is an R-morphism onto $\ker(\phi)e$. Thus for some R-submodule N of $(S^e)e$

$$(S^e)e = N \oplus \ker(\phi)e \ .$$

Now $\ker(S^e \to (S/mS)^e) = m(S^e)$. But the image of $\ker(\phi)e$ in $(S/mS)^e$ is zero. Thus $\ker(\phi)e \subseteq m(S^e)$. Hence

$$\ker(\phi)e \subseteq m(S^e)e = mN \oplus m\ker(\phi)e$$

and thus

$$\ker(\phi)e = m\ker(\phi)e \ .$$

By Nakayama's Lemma (V.2) $\ker(\phi)e = 0$. That is, e is a separability idempotent for S . The result follows by (XIV.1)(c).

(XIV.4) LEMMA. Let S be a separable extension of R . If N is an S-module which is R-free then N is S-free.

Proof. Suppose that N is a free R-module and $P \xrightarrow{\eta} N \to 0$ is an exact sequence of S-modules. As an exact sequence of R-modules the sequence splits with splitting R-morphism $\psi : N \to P$. Suppose $e = \sum x_i \otimes y_i$ is a separability idempotent for S . Define $\psi^* : N \to P$ by $\psi^*(n) = \sum x_i \psi(y_i n)$. One checks that ψ^* is a splitting S-morphism for $P \xrightarrow{\eta} N \to 0$. Hence N is S-projective. Since S is local by (V.2) N is S-free.

(XIV.5) LEMMA.

(a) Let S be a separable extension of R. If R is a finite field then S is a finite field.

(b) If K is a finite field containing a finite field k, then K is a separable extension of k.

Proof. To show (a) suppose that N is an S-module. Then N is also an R-module and if R is a field then N is R-free. By (XIV.4) N is S-free. That is, S is a local ring such that every S-module is S-free. Consequently, S is a finite field.

To show (b) we construct a separability idempotent for K. Let $|K| = p^n$ and K* denote the cyclic group of units of K. Note $|K^*| = t = p^n - 1$ thus $(t, p^n) = 1$. Suppose $K^* = \langle v \rangle$. Since t is a unit in K consider

$$e = \frac{1}{t} \sum_{i=0}^{t} v^i \otimes v^{t-i} .$$

Clearly, under $\phi : K \otimes_k K \to K$, $\phi(e) = 1$. Further, for s in k

$$\begin{aligned} (s \otimes 1)e &= \frac{1}{t} \sum_{i=0}^{t} sv^i \otimes v^{t-i} \\ &= \frac{1}{t} \sum_{\sigma} \sigma \otimes \sigma^{-1}s \\ &= (1 \otimes s)e \end{aligned}$$

where the sum $\sum_{\sigma}$ extends over all σ in K*. Thus e is a separability idempotent and K is k-separable.

(XIV.6) THEOREM. The following are equivalent.

(a) S is a separable extension of R.

(b) S is an unramified extension of R.

Proof. Suppose S is a separable extension of R. By (XIV.3) S/mS is a separable extension of R/m. Since S/mS is local (homomorphic images of local rings are local) by (XIV.5)(a) S/mS is a finite field. Then mS is a maximal ideal; i.e., $mS = M$.

Conversely, if S is an unramified extension of R; that is, $mS = M$ then $S/mS = S/M = K$ is a finite field containing $R/m = k$. By (XIV.5)(b) K is k-separable. Thus, (XIV.3) implies S is R-separable.

(XIV.7) THEOREM. (Primitive Element Theorem). Let S be a separable extension of R. Then S is a simple extension of R; i.e., $S = R[a]$ for some a in S.

Proof. By (XIV.6) $M = mS$. Since S/mS is a finite field it has a cyclic group of units with generator $\bar{a}$. Let a be a pre-image of $\bar{a}$ in S. Since $S/mS = (R/m)[\bar{a}]$ we have $S = R[a] + mS$. By Nakayama's Lemma (V.2) $S = R[a]$.

Theorem (XIV.7) is extremely important in our development of a Galois theory for commutative local rings in that it permits the introduction of polynomial theory.

(XIV.8) THEOREM. (Characterization of Separable Extensions). The local ring S is a separable extension of R if and only if $S \simeq R[X]/(f)$ (as an R-algebra) where f is a monic basic irreducible.

Proof. Suppose that S is a separable extension of R. Recall $S/M = K$, $R/m = k$ and $mS = M$. The field K is a simple extension of k. Thus let $K = k[\bar{a}]$ and let $\bar{f}$ be the minimal polynomial of $\bar{a}$ in $k[X]$ ($\bar{f}$ is monic and irreducible). Let h be a monic pre-image of $\bar{f}$ in $R[X]$ with $n = \deg(h) = [K:k]$. Then if a is a pre-image of $\bar{a}$, since $S = R[a]$, we have that $h(a)$ may be written as $\sum_{i=0}^{n-1} m_i a^i$ where the m_i are in m. Set $g(X) = \sum_{i=0}^{n-1} m_i X^i$ and $f(X) = h(X) - g(X)$. Then $f(a) = 0$ and f is monic with $\mu f = \bar{f}$; consequently, f is a basic irreducible.

Since $f(a) = 0$ we have a natural ring morphism $R[X] \to S$ by $X \to a$ which induces a surjective R-algebra morphism $\sigma : R[X]/(f) \to S$.

We claim that $|S| = |R[X]/(f)|$. If this is so then σ is also injective and thus σ is an isomorphism.

Note $|R[X]/(f)| = |R|^{\deg(f)} = |R|^{[K:k]}$. Let β denote the nilpotency of M. We have the following natural sequence of surjective ring morphisms σ_i

$$S = S/M^{\beta} \xrightarrow{\sigma_\beta} S/M^{\beta-1} \xrightarrow{\sigma_{\beta-1}} \cdots \xrightarrow{\sigma_2} S/M = K$$

where $\ker(\sigma_i) = M^{i-1}/M^i$. Clearly

$$|S| = |K| \prod_{i=2}^{\beta} |\ker(\sigma_i)|$$

$$= |K| \prod_{i=2}^{\beta} |M^{i-1}/M^i| .$$

Since $M = mS$ and $mS \otimes_R m^i \subseteq S \otimes_R m^i$ we have

$$M^{i-1}/M^i = Sm^{i-1}/Sm^i$$

$$\simeq (S \otimes_R m^{i-1})/(S \otimes_R m^i)$$

$$\simeq (S \otimes_R m^{i-1})/(mS \otimes_R m^i + S \otimes_R m^i)$$

$$\simeq (S/mS) \otimes_k (m^{i-1}/m^i)$$

$$\simeq K \otimes_k (m^{i-1}/m^i) .$$

Thus

$$|M^{i-1}/M^i| = |m^{i-1}/m^i|^{[K:k]}$$

and

$$|S| = |K| \left(\prod_{i=2}^{\beta} |m^{i-1}/m^i|\right)^{[K:k]}$$

$$= \left(|k| \prod_{i=2}^{\beta} |m^{i-1}/m^i|\right)^{[K:k]}$$

$$= |R|^{[K:k]} .$$

Thus σ is an R-algebra isomorphism.

Conversely, it is easy to see that if f is a monic basic irreducible then $R[X]/(f)$ is a local unramified extension of R .

A polynomial f is called <u>local</u> if $R[X]/(f)$ is a local extension of R . A regular polynomial is called <u>separable</u> if $R[X]/(f)$ is a local separable extension of R . It is clear that if f is separable then $(f) = (f^*)$ where f^* is monic and $\mu f = \mu f^*$. Thus the separable polynomials are the basic irreducibles. Precisely, the following are equivalent.

(a) f is separable.

(b) f is a basic irreducible.

(c) μf is irreducible.

<u>(XIV.9)</u> <u>THEOREM</u>. A regular polynomial is local if and only if μf is a power of an irreducible polynomial in $k[X]$.

<u>Proof</u>. If μf is not a power of an irreducible polynomial in $k[X]$ then by Hensel's Lemma f is not local. Conversely, if f is not local then $R[X]/(f)$ decomposes as a direct sum of ideals. Thus f and, consequently, μf factor into coprime polynomials.

<u>(XIV.10)</u> <u>COROLLARY</u>. If f is a regular irreducible polynomial in $R[X]$ then $R[X]/(f^n)$ is a local ring for any positive integer n .

(XIV.1) EXERCISE. (Nagahara). Show the following are equivalent for f in $R[X]$:

(a) f is separable,

(b) $f'(X + f)$ is invertible in $R[X]/(f)$

(f' denotes the formal derivative).

(XIV.2) EXERCISE. (Janusz). Let f be a monic polynomial of degree n in $R[X]$. The free R-module $R[X]/(f)$ has a natural basis $X^0 = 1, X, \ldots, X^{n-1}$. Let Π_i denote the natural projection of $R[X]/(f)$ onto the coefficient of X^i . Then the trace map $tr : R[X]/(f) \to R$ is defined by $tr(g) = \sum \Pi_i(gX^i)$ for g in $R[X]/(f)$. Show f is separable if and only if the determinant of the matrix $[tr(X^i X^j)]$, $0 \leq i,j \leq \deg(f)$, is a unit of R .

(XIV.3) EXERCISE. (Janusz). For $n > 1$ and r in R , show $X^n - r$ is separable if and only if $n \cdot 1$ and r are units in R .

(XIV.4) EXERCISE. Prove Theorem (XIV.2).

(XIV.5) EXERCISE. Let R be a local commutative ring of characteristic p^λ .

(a) Let G be a finite group with order prime to p . Show the group ring $R[G]$ is a separable R-algebra.

(b) Let V be a free R-module of finite dimension. Show $End_R(V)$ is a separable R-algebra.

XV. THE GALOIS THEORY OF LOCAL COMMUTATIVE RINGS

The purpose of this section is to describe the Galois extensions of a local commutative ring R and show that this theory is nearly as precise as that developed for finite fields in (II). The main results are summarized in (XV.11).

NOTE: Throughout this section S and R denote local commutative rings with maximal ideals M and m and residue fields $K = S/M$ and $k = R/m$, respectively.

Suppose that S is an extension of R and H is a group of R-algebra automorphisms of S. Let

$$S^H = \{s \text{ in } S \mid \sigma(s) = s \text{ for all } \sigma \text{ in } H\} .$$

We assume that all groups are finite unless otherwise specified. Recall if H is a subgroup of G then [G:H] denotes the index of H in G and $|H|$ denotes the order of H . If G is a group of R-algebra automorphisms (briefly, "R-automorphisms") of S then S is a <u>Galois extension</u> of R with <u>Galois</u> group G, denoted $G_R(S)$, if

(a) $S^G = R$

(b) S is a separable extension of R.

The following lemma will provide the basic step on which the theory of Galois extensions is based.

(XV.1) LEMMA. Let f be a regular polynomial in $R[X]$ and suppose μf has a simple, i.e., non-multiple, zero $\bar{\alpha}$ in k. Then f has one and only one zero α with $\mu\alpha = \bar{\alpha}$.

Proof. Let $\mu f = (X - \bar{\alpha})\bar{h}$. Since $\bar{\alpha}$ is a zero of multiplicity one, $X - \bar{\alpha}$ and $\bar{h}$ are coprime. By Hensel's Lemma

$$f = (X - \alpha + g_1)(h + g_2)$$

where g_1 and g_2 are in $m[X]$ and $\mu h = \bar{h}$. Suppose $g_1 = a_nX^n + \cdots + a_1X + a_0$ where a_i are in m. Then

$$X - \alpha + g_1 = a_nX^n + \cdots + a_2X^2 + (a_1 + 1)X + (a_0 - \alpha) .$$

By the proof of (XIII.6) there is a unit e in $R[X]$ such that

$$(X - \alpha + g_1) = e(X - \beta)$$

where β is in R and $\mu\beta = \bar{\alpha} = \mu\alpha$. Thus

$$f = e(X - \beta)(h + g_2)$$

and β is the desired zero.

Suppose β' is a second zero of f with $\mu\beta' = \bar{\alpha}$. Since $f = (X - \beta)g$ where $g = e(h + g_2)$ we have

$$0 = f(\beta') = (\beta' - \beta)g(\beta') .$$

But $\mu[g(\beta')] = \bar{h}(\bar{\alpha}) \neq 0$ since $\bar{\alpha}$ is a simple zero. Thus $g(\beta')$ is a unit and $\beta = \beta'$.

(XV.2) THEOREM. (The Lifting Theorem). Let S be a separable extension of R. Let T be a local commutative extension of R with residue field K. Then for each k-algebra isomorphism $\bar{\sigma} : K \to K$ there exists one and only one R-algebra morphism $\sigma : S \to T$ which induces $\bar{\sigma}$ modulo the maximal ideals. Further σ is an R-algebra isomorphism if and only if T is a separable extension of R.

Proof. Let $K = k[\bar{a}]$ where $\bar{a}$ is a zero of $\bar{f} = \mathrm{Irr}(k,\bar{a})$. Since $\bar{a}$ is a simple zero of $\bar{f}$, if f is a pre-image of $\bar{f}$ in $R[X]$ then by (XV.1) there is precisely one element a in S with $f(a) = 0$ and S is R-free on $\{1,a,\ldots,a^{n-1}\}$ where $n = \deg(\bar{f})$. Indeed, $S \simeq R[X]/(f)$.

Let $\bar{\sigma} : K \to K$ be a k-algebra isomorphism and suppose $\bar{\sigma}(\bar{a}) = \bar{a}_0$. Then $K = k[\bar{a}_0]$ and $\bar{a}_0$ is also a zero of $\bar{f}$.

As above f has precisely one zero a_0 in T which is a pre-image of $\bar{a}_0$. The map $a \to a_0$ extends naturally to an R-algebra morphism $\sigma : S \to T$. Further σ induces $\bar{\sigma}$.

Conversely, if $\sigma : S \to T$ is an R-algebra morphism which induces $\bar{\sigma} : \bar{a} \to \bar{a}_0$ on K then σ maps a to a zero of f , say b_0 . By the uniqueness (since $\mu b_0 = \mu a_0$) of (XV.1) $a_0 = b_0$.

The last statement is clear.

<u>(XV.3)</u> <u>COROLLARY</u>. Let S be a separable extension of R . Then S is a Galois extension of R with group $G_R(S)$ isomorphic to $G_k(K)$ (thus $G_R(S)$ is a finite cyclic group).

<u>Proof</u>. Since S is a separable extension of R by (XV.2) each $\bar{\sigma}$ in $G_k(K)$ may be lifted to precisely one R-automorphism σ of S . Further every R-automorphism of S is obtained in this fashion. Let G denote the group of R-automorphisms of S . It remains to show that $R = S^G$. Obviously $R \subseteq S^G$. Suppose s is in S and s is not in R . If s is a unit then μs is moved by some $\bar{\sigma}$ in $G_k(K)$. Thus, if σ induces $\bar{\sigma}$, $\sigma(s) \neq s$. If s is a non-unit, then $1 + s$ is a unit and $1 + s$ is not in R . As above there is a σ in G with $\sigma(1 + s) = 1 + \sigma(s) \neq 1 + s$; i.e., $\sigma(s) \neq s$. Thus $R = S^G$.

Since a Galois extension S of R is certainly separable we have the following from (XIV.6).

<u>(XV.4)</u> <u>COROLLARY</u>. The following are equivalent for an extension S of R.

(a) S is a Galois extension of R.

(b) S is a separable extension of R.

(c) S is an unramified extension of R.

Using the construction in the proof of (XV.2) and (XIV.8),

<u>(XV.5)</u> <u>COROLLARY</u>. Let S be a Galois extension of R. Then $S \simeq R[X]/(f)$ where f is a basic irreducible (which we may assume is monic) and $G_R(S)$ permutes the zeros of f.

We note that if $S \simeq R[X]/(f)$ where f is a monic basic irreducible then S is R-free and

$$\deg(f) = \dim_R(S) = |G_R(S)| = |G_k(K)| = \dim_k(K) = \deg(\mu f) .$$

<u>(XV.6)</u> <u>THEOREM</u>. Let S be a Galois extension of R with $\dim_R(S) = n$. Let t denote the number of monic basic irreducible polynomials in $R[X]$ of degree n and let r denote the number of primitive elements of S over R. Then $tn = r$.

Proof. Note that $t = \bar{t}|m|^n$ and $r = \bar{r}|M|$ where $\bar{t}$ and $\bar{r}$ are the analogous values for the field extension $K = S/M$ over $k = R/m$. Since $\bar{t}[K:k] = \bar{t}n = \bar{r}$ it is only necessary to show $|m|^n = |M|$. If a is a primitive element for S then $\{1,a,\ldots,a^{n-1}\}$ is an R-basis for S and $\{1,\bar{a},\ldots,\bar{a}^{n-1}\}$ where $\mu a = \bar{a}$ is a k-basis for K. If $d_0 + d_1a + \cdots + d_{n-1}a^{n-1}$ is in M where the d_i are in R then $\bar{d}_0 + \cdots + \bar{d}_{n-1}\bar{a}^{n-1} = 0$ where $\bar{d}_i = \mu d_i$. Thus, $\bar{d}_i = 0$, $0 \leq i \leq n-1$, and the d_i are in $R \cap M = m$. Conversely, any element of the above form is in M. Hence $|M| = |m|^n$.

(XV.7) THEOREM. (Uniqueness of Galois Extension). Let T be a local commutative ring. Let $R \subseteq S \subseteq T$. If S is a Galois extension of R with $\dim_R(S) = n$ then S is the only Galois extension of R in T of dimension n.

Proof. It is a standard argument to show the uniqueness of S is equivalent to (XV.6).

An element a of S is called R-separable or separable if a is a zero of a basic irreducible polynomial in $R[X]$.

Observe if $R \subseteq T \subseteq S$ and S is a Galois extension of R then T is a separable, hence Galois, extension of R if and only if $T/T \cap M$ is a subfield of K containing k. That is, there is a lattice preserving bijection between the subfields k_1, $k \subseteq k_1 \subseteq k$, and the separable extensions R_1 of R , $R \subseteq R_1 \subseteq S$.

EXAMPLE. Let $R = Z/4Z$ and $f = X^3 + X + 1$ in $R[X]$. Since μf is irreducible in $Z/2Z$, f is a basic irreducible. Hence $S = R[X]/(f)$ is a Galois extension of R . Since $\deg(f) = 3$, $\dim_R(S) = 3$ and $|G_R(S)| = 3$; thus $G_R(S)$ is cyclic of order 3 and there are no proper R-separable subrings of S .

EXAMPLE. Let $R = Z/4Z$ and $f = X^4 + X^3 + X^2 + X + 1$ be in $R[X]$. Since f is irreducible, $S = R[X]/(f)$ is a Galois extension of R with $G_R(S) = \langle\sigma\rangle$ cyclic of order 4. There is the following chain of separable extensions of R : $R \subseteq T \subseteq S$ where $T \simeq R[X]/(g)$ and $g = X^2 + X + 1$. By the Lifting Theorem we may construct the R-automorphisms of S from $G_k(K)$. If a is a root of f in S then

$$f = (X - a)(X - a^2)(X - a^3)(X - [3a^3 + 3a^2 + 3a + 3]) .$$

The set $\{1,a,a^2,a^3\}$ is an R-basis for S . Then

$$\sigma a = a^2$$

$$\sigma^2 a = 3a^3 + 3a^2 + 3a + 3$$

$$\sigma^3 a = a^3$$

$$\sigma^4 a = a$$

defines σ,σ^2,σ^3 and σ^4 = identity . The polynomial

$$g = [X - (a^3 + a^2 + 2)][X - (3a^3 + 3a^2 + 1)] .$$

Observe

$$\sigma^2(a^3 + a^2 + 2) = a^3 + a^2 + 2$$

so $a^3 + a^2 + 2$ is in the fixed ring of $\langle\sigma^2\rangle$ and $T = R[a^3 + a^2 + 2]$.

Although for commutative local rings the Galois group is cyclic the generating automorphism of the Galois group cannot be described as a power map on all the elements as in (II.6).

EXAMPLE. Let $R = Z/4Z$ and $S = R[a]$ where a satisfies $X^2 + X + 1$. Then $G_R(S) = \{1,\sigma\}$ where $\sigma a = 3a + 3$. Note that no power of $2a$ is $2a + 2 = \sigma(2a)$. Also $\sigma(3a + 1) = a + 2$ and it is easy to check that $a + 2$ is not a power of $3a + 1$. However, $\rho a = 3a + 3 = a^2$. Thus $G_R(S)$ is generated by an automorphism which takes a primitive element to its square.

We now show that $G_R(S)$ may be described as a power map on certain primitive elements.

(XV.8) LEMMA. Let S be a Galois extension of R . Let $S = R[a]$ and $a = a_1,\ldots,a_n$ be the distinct images of a under $G_R(S)$. If g is in $R[X]$ and $g(a) = 0$ then g is a multiple of $f = (X - a_1) \cdots (X - a_n)$.

Proof. Note that $a_i - a_j$ is a unit for $i \neq j$. If $g(a) = 0$ then $g(a_i) = 0$ for each i . We can find a p_1 in $S[X]$ with $g = (X - a_1)p_1$. Since $g(a_2) = 0$ and $a_2 - a_1$ is a unit, $p_1(a_2) = 0$. Then there is a p_2 in $S[X]$ with $p_1 = (X - a_2)p_2$. Continuing, we obtain $g = fp_n$. Since f and g are in $R[X]$ and f is monic, p is in $R[X]$.

(XV.9) LEMMA. Let S be a Galois extension of R . Let f be a monic basic irreducible in $R[X]$. If a and b are zeros of f in S then there exists a monic basic irreducible g in $R[X]$ such that $a^{|k|}$ and $b^{|k|}$ are zeros of g .

Proof. It is easy to see that we may select a monic basic irreducible g in $R[X]$ with $g(a^{|k|}) = 0$ and $\mu g = \mu f$. Consider the polynomial $h(X) = g(X^{|k|})$. Clearly $h(a) = 0$. But by (XV.8) f divides h . Thus $h(b) = 0$; i.e., $b^{|k|}$ is a zero of g .

(XV.10) THEOREM. Let S be a Galois extension of R . Then there exists a primitive element a of S over R such that the R-automorphism σ of S given by $\sigma : a \to a^{|k|}$ is a generator of $G_R(S)$.

Proof. Let f be a monic basic irreducible of degree n in $R[X]$ and let a be a zero of f . Set $A = \{g$ in $R[X] \mid g$ is monic and $\mu g = \mu f\}$, $B = \{b$ in $S \mid b$ is a root of some polynomial in $A\}$ and $B^j = \{b^j \mid b$ in $B\}$ for $j = 1,2,\ldots$. Clearly $B \supseteq B^{|k|} \supseteq B^{|k|^2} \supseteq \cdots$. If $\bar{a} = \mu a$ since $\mu f = (X - \bar{a})(X - \bar{a}^{|k|})\cdots(X - \bar{a}^{|k|^{n-1}})$ and each element of B is a pre-image of one of $\{\bar{a},\bar{a}^{|k|},\ldots,\bar{a}^{|k|^{n-1}}\}$, then each element of B has the form $a^j + c$ where c is in M and $j = 1,|k|,\ldots,|k$ Hence, since M is nilpotent, there exists an s such that $B^s = B^{s+1} = \cdots$ and B^s has only n elements.

Raising each element of B^s to the $|k|^{th}$-power results only in reshuffling By the Lifting Theorem and (XV.9) there is an R-automorphism σ of S given by $\sigma t = t^{|k|}$ where t is in B^s . Consider the induced k-automorphisms $\bar{\sigma},\bar{\sigma}^2,\ldots,\bar{\sigma}^n$ in $G_k(K)$. These k-automorphisms are distinct since the map $\mu t \to (\mu t)^{|k|}$ generates $G_k(K)$. Thus σ generates $G_R(S)$ and we are done.

A Galois extension S of R is called a <u>splitting ring</u> for a basic irreducible f in R[X] if f is a product of linear factors in S[X] and S is generated over R by the zeros of f .

We summarize the results of this section with the following theorem. Those parts which we have not shown are immediate.

<u>(XV.11)</u> <u>THEOREM</u>. Let S be a separable extension of R . Then

(a) S is a Galois extension of R and if f is a monic basic irreducible such that $S \simeq R[X]/(f)$, then $|G_R(S)| = \deg(f)$, S is a splitting ring of f and S is the unique Galois extension of R having R-dimension equal to deg(f) .

(b) $G_R(S)$ is cyclic, isomorphic to $G_k(K)$ and generated by a power map

$$\sigma : a \to a^{|k|}$$

on a suitable primitive element a .

(c) There is a lattice preserving bijection between the subfields of K containing k and the R-separable subrings of S containing R . If T is a separable extension of R in S then S is a separable extension of T and

$$1 \to G_R(T) \to G_R(S) \to G_T(S) \to 1$$

is exact.

(d) S has a normal basis over R ; i.e., there exists an element a in S such that $\{\sigma a \mid \sigma \text{ in } G_R(S)\}$ is a free R-basis for S .

(XV.1) EXERCISE. Let $\nu(n,R)$ denote the number of monic basic irreducible polynomials of degree n in $R[X]$. If $n = p_1^{e_1} \cdots p_s^{e_s}$ show (recall (II.3)) that

$$\begin{aligned}\nu(n,R) &= \nu(n,k)|m|^n \\ &= \frac{1}{n}|R|^n (1 - \sum |k|^{1/p_i} + \sum |k|^{1/p_ip_i} - \cdots \\ &\qquad + (-1)^s |k|^{1/p_1\cdots p_s}) .\end{aligned}$$

If S is a Galois extension of R with $\dim_R(S) = n$ show there exist $n\nu(n,R)$ primitive elements for S over R.

<u>(XV.2)</u> <u>EXERCISE</u>. Show the local case provides information about the automorphism group of any finite commutative ring in the following fashion. Suppose $\bar{S} = \oplus \sum_{i=1}^{n} S_i$ where S_i is a local extension of a local ring R_i. Let e_i be the identity of S_i and $\bar{R} = \oplus \sum_{i=1}^{n} R_i$. Let $\Pi_i : \bar{S} \to S_i$ be the natural projection and $\lambda_i : S_i \to \bar{S}$ be the natural injection. For a in S_i let $a^{(i)} = \lambda_i a$. If σ is an $\bar{R}$-automorphism of $\bar{S}$ then $\sigma a^{(i)} = \sigma(e_i^{(i)}a)$. But $e_i^{(i)}$ is in $\bar{R}$ thus $\sigma a^{(i)} = e_i^{(i)}\sigma(a^{(i)})$ is in S_i and since σ is an isomorphism $\sigma S_i = S_i$. Set $\sigma_i = \Pi_i \sigma \lambda_i$. Then σ_i is an R_i-automorphism of S_i and $\sigma = \sigma_1 \oplus \cdots \oplus \sigma_n$. Further $\bar{S}$ is separable over $\bar{R}$ if and only if for each i, $1 \leq i \leq n$, S_i is separable over R_i.

<u>(XV.3)</u> <u>EXERCISE</u>. Show that $X^2 - 2$ has 0 as a multiple zero in $Z/Z2$ but has no zero in $Z/8Z$ (see Lemma (XV.1)).

<u>(XV.4)</u> <u>EXERCISE</u>. Let R be a local ring and G be a group. Let $\mathrm{Map}(G,R)$ be the R-algebra of all maps $\sigma : G \to R$. A map $\sigma : G \to R$ is called an <u>R-character</u> if $\sigma(G) \subseteq R^*$ and σ is a group morphism. Show the set of R-characters is R-free in $\mathrm{Map}(G,R)$. (Hint: Use induction on the sets of R-free characters.)

<u>(XV.5)</u> <u>EXERCISE</u>. Let $R \subseteq S$ be local rings. The mapping $\Phi : S \to \text{End}_R(S)$ by $\phi(s)(t) = st$ for all t in S is an R-algebra morphism. Let G be a group of R-algebra automorphisms of S. Then G may be considered as a subgroup of $GL_R(S) \subseteq \text{End}_R(S)$. We raise the question, "What is the R-algebra generated by G and $\Phi(S)$ in $\text{End}_R(S)$?" Show the following are equivalent. (See Section II (D).)

(a) S is a Galois extension of R and $G = G_R(S)$.

(b) S is R-free and $\text{End}_R(S)$ is generated by $\Phi(S)$ and G.

The above may be formulated as follows. Let G be a group of R-algebra automorphisms of S. Let $\Delta(S:G)$ be a free S-module on $\{u_\sigma \mid \sigma \text{ in } G\}$. Define multiplication on $\Delta(S:G)$ by $(su_\sigma)(tu_\beta) = s\sigma(t)u_{\sigma\beta}$ and extend linearly. Define

$$\phi : \Delta(S:G) \to \text{End}_R(S)$$

by $\phi(su_\sigma)(t) = s\sigma(t)$. Show the following are equivalent.

(a) S is a Galois extension of R and $G = G_R(S)$.

(b) S is R-free and $\phi : \Delta(S:G) \to \text{End}_R(S)$ is an isomorphism.

The above is a starting point to the Galois theory of commutative rings and R-algebras (see Chase, Harrison, Rosenberg [7]; Ingraham, DeMeyer [14]). The next exercise provides an alternate approach.

<u>(XV.6)</u> <u>EXERCISE</u>. Let $R \subseteq S$ be local rings. Let G be a group of R-algebra automorphisms of S. Let $\nabla(S:G)$ be a free S-module on $\{v_\sigma \mid \sigma \text{ in } G\}$. Define multiplication "coordinate-wise", i.e.,

$$(sv_\sigma)(tv_\beta) = \begin{cases} stv_\sigma & \text{if } \sigma = \beta \\ 0 & \text{otherwise} \end{cases} .$$

Define

$$\beta : S \otimes_R S \to \nabla(S:G)$$

by

$$\beta(s \otimes t) = \sum_{\sigma \text{ in } G} s\sigma(t)v_\sigma .$$

Show the following are equivalent.

(a) S is a Galois extension of R with $G = G_R(S)$.

(b) $S^G = R$ and $\beta : S \otimes_R S \to \nabla(S:G)$ is an isomorphism.

(Hint: see the above references.)

<u>(XV.7)</u> <u>EXERCISE</u>. Let $r_1,\ldots,r_m$ be distinct units in the local ring R . $a_1,\ldots,a_m$ are in R and

$$a_1r_1^n + \cdots + a_mr_m^n = 0$$

for all integers n , show $a_i = 0$, $1 \le i \le m$. (Hint: Apply the above exercise on characters with $G = Z$ and $\sigma_i(n) = a_i^n$.)

<u>(XV.8)</u> <u>EXERCISE</u>. Let $R \subseteq S$ be local rings with S a Galois extension of R . For s in S , define $\Phi : S \to \mathrm{End}_R(S)$ by $\Phi(s)(t) = st$. Since S is R-free, $\Phi(s)$ has a matrix. Define

$$N_{S/R}(s) = \text{determinant of } \Phi(s)$$

$$T_{S/R}(s) = \text{trace of } \Phi(s) .$$

(a) Let $\langle\sigma\rangle = G_R(S)$. Let β be in S . Show $N_{S/R}(\beta) = 1$ if and only if there is a unit α in S with $\beta = \alpha\sigma(\alpha)^{-1}$.

(b) Let $\langle\sigma\rangle = G_R(S)$. Let β be in S . Show $T_{S/R}(\beta) = 0$ if and only if there is an α in S satisfying $\beta = \alpha - \sigma(\alpha)$. (Hint: See any text on field theory and examine Hilbert's Theorem 90.)

(XV.9) EXERCISE. Prove Theorem (XV.7).

(XV.10) EXERCISE. Prove Theorem (XV.11).

(XV.12) EXERCISE. (Menzin [35]). Let R be a local ring and M a finitely generated R-module. Let $L_R(M)$ denote the length of M (see Exercise (III.3)). Let S be a Galois extension of R . Show

(a) If $0 \to N \to M \to M/N \to 0$ is an exact sequence of R-modules, then

$$0 \to S \otimes_R N \to S \otimes_R M \to S \otimes_R (M/N) \to 0$$

is an exact sequence of S-modules.

(b) $L_R(M) = L_S(S \otimes_R M)$.

(c) Determine $S \otimes_R S$. (See (VIII.9) and (XVI.8).)

(d) Let $n = [S:R]$ and N be a faithful indecomposable finitely generated S-module. Let $N = \oplus \sum_{i=1}^{t} N_i$ be a decomposition of of N into indecomposable R-modules. Then

(1) $t \leq n$

(2) $L_S(N) \leq \max \{L_R(N_i)\}$

(3) The N_i are R-faithful.

(If R is not principal then the above may be used to construct faithful indecomposable R-modules of arbitrarily large length. See Menzin [35] for hints and additional remarks.)

XVI. GALOIS RINGS

In this section we examine Galois Rings - Galois extensions of the prime ring Z/Zp^n. The next few pages characterize finitely generated modules over a Galois ring, the tensor product of Galois rings, and the group of units of a Galois ring.

The value of these rings are two-fold. First, it is classically accepted that in certain classes of problems in combinatorial mathematics the researcher handles separately the finite field $GF(p^n)$ and the prime ring Z/Zp^m. It is our belief that both cases should be treated simultaneously in the setting of a Galois ring. Second, we show shortly that any finite ring may be naturally regarded as a faithful algebra over a Galois ring.

This section is devoted to a study of the Galois extensions of the prime ring Z/Zp^n, modules over these rings and their group of units. For a prime p, $GR(p^n,r)$ denotes the Galois extension of Z/Zp^n of degree

r . This extension $GR(p^n,r)$ is called a <u>Galois Ring</u>. Observe $GR(p,r) = GF(p^r)$ and $GR(p^n,1) = Z/Zp^n$.

The existence and structure for Galois rings was known to Krull in 1924 [54]. Janusz ([68], 1966) and Raghavendran ([131],1969) independently rediscovered these rings.

The previous sections tell us that $GR(p^n,r)$ is isomorphic to $(Z/Zp^n)[X]/(f)$ where f is a monic basic irreducible in $(Z/Zp^n)[X]$ of degree r . Equivalently, if f is a monic polynomial of degree r in $Z[X]$ which is irreducible modulo pZ then $GR(p^n,r) \simeq Z[X]/(p^n,f)$.

The maximal ideal in $GR(p^n,r)$ is principal and of the form $pGR(p^n,r)$; indeed, every ideal has the form $(p^i) = p^i GR(p^n,r)$ where $0 \le i \le n$. Each non-zero element in $GR(p^n,r)$ may be written as up^t where u is a unit and $0 \le t \le n - 1$. It is easy to check that in the representation of an element as up^t the integer t is unique and u is unique modulo (p^{n-t}) .

Certain mappings between the Galois rings are worth noting. For each positive integer d there is a natural inclusion of $GR(p^n,r)$ into $GR(p^n,dr)$. A second map is the natural projection $GR(p^n,r) \to GR(p^{n-1},r)$ with kernel (p^{n-1}) .

Our first goal is a study of left and two-sided $GR(p^n,r)$-modules. As noted earlier all modules are finitely generated.

Let M be a finitely generated left module over $GR(p^n,r)$. A <u>finite presentation</u> of M is an exact sequence

$$F_1 \xrightarrow{\rho} F_0 \xrightarrow{\sigma} M \longrightarrow 0$$

of $GR(p^n,r)$-modules (thus $\ker(\sigma) = \mathrm{Im}(\rho)$ and $\mathrm{Im}(\sigma) = M$) and $GR(p^n,r)$-morphisms where F_0 and F_1 are free finitely generated $GR(p^n,r)$-modules.

<u>(XVI.1)</u> <u>LEMMA</u>. Let M be a finitely generated left $GR(p^n,r)$-module. Then M admits a finite presentation

$$F_1 \xrightarrow{\rho} F_0 \xrightarrow{\sigma} M \longrightarrow 0 .$$

Further, if M has a minimal generating set of s elements then it may be assumed that $s = \dim(F_0) = \dim(F_1)$.

<u>Proof</u>. Let M have a minimal generating set of s elements, say $\{m_1,\ldots,m_s\}$. Let F_0 be $GR(p^n,r)$-free on $\{x_1,\ldots,x_s\}$ and define σ by $\sigma : x_i \to m_i$. The kernel of σ is a submodule of F_0 and easily seen (use (V.5)) to be generated by $t \le s$ elements, say $\{p_1,\ldots,p_t\}$. Let F_1 be free on $\{y_1,\ldots,y_s\}$ and σ be given by

$$\sigma : y_i \to p_i , \quad 1 \le i \le t$$

$$\sigma : y_i \to 0 \ , \quad t < i \le s \ .$$

Suppose $F_1 \xrightarrow{\rho} F_0 \xrightarrow{\sigma} M \longrightarrow 0$ is a finite presentation of M as described in the above proof. Then the matrix of ρ, $\mathrm{Mat}(\rho) = [a_{ji}]$, given by $\rho(y_i) = \sum_j a_{ji}x_j$ for $1 \le i \le s$, is called a _presentation matrix of M_. Observe finally that

$$M = \mathrm{Im}(\sigma) \simeq F_0/\ker(\sigma) = F_0/\mathrm{Im}(\rho) \ .$$

(XVI.2) THEOREM. Let R denote the Galois ring $GR(p^n,r)$ and let M be a finitely generated left R-module. Then M is R-isomorphic to a direct sum of cyclic modules R/Rd_i, $1 \le i \le s$, where d_i is not a unit and d_i divides d_{i+1}, $1 \le i \le s - 1$. Further, the ideal $Rd_i = Rp^{j(i)}$ for some $j(i)$, $0 < j(i) \le n$, s is unique, and the ideals are unique up to R-isomorphism.

Proof. We only sketch the proof since it is standard. Form a finite presentation

$$F_1 \xrightarrow{\rho} F_0 \xrightarrow{\sigma} M \longrightarrow 0$$

of M with $s \times s$ presentation matrix $\mathrm{Mat}(\rho) = [a_{ji}]$.

Then (see Exercises in this section) $\mathrm{Mat}(\rho) = [a_{ji}]$ admits diagonal reduction; i.e., there exist invertible matrices P and Q such that

$$PAQ = \mathrm{diag}[d_1,d_2,\ldots,d_s]$$

such that d_i divides d_{i+1} , $1 \leq i \leq s - 1$. This is equivalent to saying that under a change of basis we may select bases $\{y_1,\ldots,y_s\}$ and $\{x_1,\ldots,x_s\}$ of F_1 and F_0 , respectively, such that

$$\rho : y_i \to d_i x_i \ .$$

Since $M \simeq F_0/\mathrm{Im}(\rho)$, we have that $M \simeq \oplus \sum_{i=1}^{n} R/Rd_i$. If $d_1,\ldots,d_j$ are units, then $R/Rd_i = 0$, $1 \leq i \leq j$; thus, without loss, we may assume d_1 is a non-unit. Further, by the previous comments, each ideal Rd_i is of the form Rp^j for some j . The uniqueness of the decomposition follows from the Krull-Schmidt Theorem since it is easy to see that R/Rd_i are indecomposable.

Let M be a finitely generated left R-module where $R = GR(p^n,r)$. We now determine the projective cover and injective envelope (to be defined below) of M . Since R is local, projective R-modules are free.

The proof of (XII.11) indicated how a projective cover of M may be constructed. Let M have a minimal R-generating set $\{m_1,\ldots,m_s\}$. Then the R/Rp-space $M/(p)M$ has a basis $\{\bar{m}_1,\ldots,\bar{m}_s\}$ where $\bar{m}_i$ is the image of m_i . Let F be a free R-module on $\{x_1,\ldots,x_s\}$ and consider

$$\begin{array}{ccccc} & & F & & \\ & & \downarrow & & \\ M & \longrightarrow & M/(p)M & \longrightarrow & 0 \end{array}$$

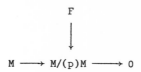

where $x_i \to \bar{m}_i$ for $1 \le i \le s$. Since F is free the map $F \to M/(p)M$ lifts to a surjective map $F \to M$ by $x_i \to m_i$ and the proof of (XII.11) indicates that F is a projective cover for M .

To make this more explicit we may utilize (XVI.2) and suppose that $M = Rm_1 \oplus \cdots \oplus Rm_s$ is a direct sum of cyclic R-modules where $\mathrm{Annih}_R(m_i) = (p^{e_i})$ for $1 \le e_i \le n$ and $1 \le i \le s$. Then $F \to M$ by $x_i \to m_i$ and $\ker(F \to M) = (p^{e_1})x_1 \oplus \cdots \oplus (p^{e_s})x_s$ is small in F .

The exercises in (XII) indicate that $R = GR(p^n,r)$ is <u>self-injective</u> or <u>quasi-Frobenius</u>, that is, ${}_RR$ is an injective R-module. Since a finite direct sum of injective modules is again injective, a free finitely generated R-module is injective.

If R is any ring the exercises in (XII) show that an R-module may be embedded in an injective R-module. If M and N are finitely generated R-modules, and $M \subseteq N$, i.e., N is an <u>extension</u> of M , then N is called an <u>injective envelope</u> of M if

(a) N is R-injective;

(b) if N' is an R-module with $M \subseteq N' \subseteq N$ and $N' \neq N$ then N' is not R-injective.

Let $E(M)$ denote an injective envelope of M . It is well-known that injective envelopes exist and that they are unique up to module isomorphism. Further, if $M_1,\ldots,M_s$ are finitely generated R-modules then it is easy to see that $E(M_1 \oplus \cdots \oplus M_s) = E(M_1) \oplus \cdots \oplus E(M_s)$.

The last statement together with (XVI.2) permits us to examine cyclic R-modules when R is a Galois ring in determining injective envelopes for finitely generated R-modules.

Let $R = GR(p^n,r)$ and $T = Rb$ be a cyclic R-module with $\text{Annih}_R(b) = (p^t)$. Let $F = Rx$ be a free R-module on $\{x\}$. There is a natural injective R-morphism $T \to F$ by $b \to p^{n-t}x$. Thus, this gives an embedding of T in an injective R-module. To show that R is the injective envelope we must show there is no cyclic R-module T_1 with $T \subsetneq T_1 \subsetneq R$ and T_1 injective. Equivalently, we are done if we show that whenever $t < n$ then T is not R-injective. But

$$0 \to T \to F = Rx$$

by $b \to p^{n-t}x$ does not split if $t < n$ since F is an indecomposable R-module. Hence by (XII.12), T is not R-injective.

<u>(XVI.3)</u> <u>THEOREM</u>. Let $R = GR(p^n,r)$ be a Galois ring and M a finitely generated R-module.

(a) The following are equivalent.

(1) M is R-free.

(2) M is R-projective.

(3) M is R-injective.

(b) The projective cover and injective envelope of M are

isomorphic to the free R-module $R^{(n)}$ where $n = \dim_{R/Rp}(M/(p)M)$

Theorem (VII.1) (together with Exercise (VII.2)) indicates how one might compute the endomorphism ring of a finitely generated module over a Galois ring. The submodules of a finitely generated module are examined in the exercises. We now examine the dual module.

Let R be a Galois ring and M a finitely generated R-module. Let $M^* = \mathrm{Hom}_R(M,R)$. The module M^* is called the <u>dual</u> of M and is naturally an R-module.

Define a <u>dual pairing</u>

$$\langle , \rangle : M \times M^* \to R$$

by

$$\langle m, m^* \rangle = m^*(m) \; .$$

The mapping $\langle , \rangle$ is R-bilinear. The proof of the following theorem is left as an exercise.

<u>(XVI.4)</u> <u>THEOREM</u>. (For the above setting.) Let Φ denote the set of submodules of M and Λ denote the set of submodules of M^* . Define

$$\lambda : \Phi \to \Lambda$$

by $N \to N^{\lambda} = \{m^* \text{ in } M^* \mid \langle m, m^* \rangle = 0 \text{ for all } m \text{ in } N\}$ and

$$\phi : \Lambda \to \Phi$$

by $P \to P^{\phi} = \{m \text{ in } M \mid \langle m,m^*\rangle = 0 \text{ for all } m^* \text{ in } P\}$. Then

(a) λ is a bijection

(b) If $N_1 \subseteq N_2$ are submodules of M , then

(1) $N_2^{\lambda} \subseteq N_1^{\lambda}$

(2) $N_1^{\lambda\phi} = N_1$

(c) If N_1 and N_2 are submodules of M , then

(1) $(N_1 + N_2)^{\lambda} = N_1^{\lambda} \cap N_2^{\lambda}$

(2) $(N_1 \cap N_2)^{\lambda} = N_1^{\lambda} + N_2^{\lambda}$

and similar statements are true for ϕ .

We now turn to the question of two-sided $GR(p^n,r)$-modules. This is somewhat more specialized. For a commutative ring R , a two-sided R-module M , denoted (R,R)-module, is a right and left R-module such that $(rm)s = r(ms)$ for r and s in R and m in M .

(XVI.5) THEOREM. Let R denote the Galois ring $GR(p^n,r)$ and let M be a two-sided R-module. Let $G_{Z/Zp^n}(R) = \{1 = \sigma_1,\sigma_2,\ldots,\sigma_r\}$. Then

$$M = M_1 \oplus \cdots \oplus M_r \qquad \text{(as } (R,R)\text{-module)}$$

where for each i there is an automorphism $\sigma_{j(i)}$ such that

$$mr = \sigma_{j(i)}(r)m$$

for every r in R and m in M_i .

Proof. Since $R = GR(p^n, r)$ is a Galois extension of Z/Zp^n, there exists a primitive element a_1 such that $R = (Z/Zp^n)[a_1]$. The element a_1 satisfies a monic basic irreducible f in $(Z/Zp^n)[X]$. Further, in $R[X]$, f splits into linear factors

$$f(X) = (X - a_1) \cdots (X - a_r) .$$

Since μa_i and μa_j, $i \neq j$, are non-zero and distinct in R/Rp (where $\mu : R \to R/Rp$), $a_i - a_j$ is a unit. Define

$$f_i = \prod_{\substack{j=1 \\ j \neq i}}^{r} (X - a_j) .$$

Then $f_i(a_i)$ is a unit and we have the following identity

$$(*) \qquad \textstyle\sum_{i=1}^{r} [f_i(a_i)]^{-1} f_i(X) = 1 .$$

Observe that right multiplication of elements in M by a_1 gives a left R-linear morphism

$$\sigma : M \to M \qquad (\sigma(m) = ma_1).$$

Let $f = X^r + b_0 X^{r-1} + \cdots + b_r$ where the b_i are in Z/Zp^n. Then

$$\begin{aligned} f(\sigma)m &= (\sigma^r + b_0\sigma^{r-1} + \cdots + b_r)(m) \\ &= ma_1^r + b_0 ma_1^{r-1} + \cdots + b_r m \\ &= mf(a_1) \\ &= 0 \end{aligned}$$

since $f(a_1) = 0$ and $b_i m = m b_i$ for b_i in Z/Zp^n. The identity (*) gives the identity mapping

$$i_M = \sum_i f_i(a_i)^{-1} f_i(\sigma) : M \to M ,$$

so that

$$M = M_1 + \cdots + M_r$$

where $M_i = f_i(a_i)^{-1} f_i(\sigma) M$ for $1 \le i \le r$.

We now show the sum $M = M_1 + \cdots + M_r$ is direct. Suppose, without loss of generality, that m is in $M_1 \cap (M_2 + \cdots + M_r)$.

Since m is in M_1, $m = f_1(a_1)^{-1} f_1(\sigma) m_1$ for some m_1 in M. Then

$$\begin{aligned}(\sigma - a_1)(m) &= f_1(a_1)^{-1}(\sigma - a_1) f_1(\sigma) m_1 \\ &= f_1(a_1)^{-1} f(\sigma) m_1 = 0 .\end{aligned}$$

On the other hand, since m is in $M_2 + \cdots + M_r$,

$$m = f_2(a_2)^{-1} f_2(\sigma) m_2 + \cdots + f_r(a_r)^{-1} f_r(\sigma) m_r$$

for some $m_2, \ldots, m_r$ in M. Then

$$(\sigma - a_2) \cdots (\sigma - a_r)(m) = \left[\prod_{i=2}^{r} (\sigma - a_i)\right]\left[\sum_{i=2}^{r} f_i(a_i)^{-1} f_i(\sigma) m_i\right] = 0 .$$

Thus the sum is direct.

For each m in M_i, $m = f_i(a_i)^{-1} f_i(\sigma) m_i$ for some m_i in M and

$$(\sigma - a_i)m = (\sigma - a_i)f_i(a_i)^{-1}f_i(\sigma)m_i$$
$$= f_i(a_i)^{-1}f(\sigma)m_i$$
$$= 0 .$$

Thus $\sigma(m) = a_i m$. Consequently, for each m in M_i ,

$$ma_1 = \sigma(m) = a_i m .$$

The elements a_i and a_1 are zeros of f thus there is an automorphism, say $\sigma_{j(i)}$, in the Galois group $G_{Z/Zp^n}(R)$ such that $\sigma_{j(i)}(a_1) = a_i$.

That is, for each m in M_i , $ma_1 = \sigma_{j(i)}(a_1)m$. Since $R = (Z/Zp^n)[a_1]$, each s in R has the form

$$s = c_0 + c_1a_1 + \cdots + c_{r-1}a_1^{r-1}$$

where c_i is in Z/Zp^n , and it is easy to see that

$$ms = \sigma_{j(i)}(s)m$$

for every m in M . This completes the proof.

Observe that the 'skew' multiplication

$$ms = \sigma(s)m$$

where σ is an automorphism as described above permits an (R,R)-module M (R a Galois ring) to be considered as a left R-module since the 'scalars' may be shifted to the left. Combining (XVI.2) and (XVI.5) we have the following result.

(XVI.6) THEOREM. Let R denote the Galois ring $GR(p^n,r)$ and let M be a two-sided R-module. Let $\{1 = \sigma_1,\sigma_2,\ldots,\sigma_r\}$ be the Z/Zp^n-automorphisms of R. Then M is the direct sum of cyclic modules

$$M = \oplus \sum_{j=1}^{r} \sum_{i=1}^{n_j} Rb_{ij}$$

where

$$b_{ij}r = \sigma_j(r)b_{ij} \ , \quad 1 \leq i \leq n_j \ .$$

We now turn to an examination of the tensor product of Galois rings. The following lemma is an easy exercise.

(XVI.7) LEMMA. Every subring of $GR(p^n,r)$ is a Galois ring of the form $GR(p^n,s)$ where s divides r. Conversely, if s divides r then $GR(p^n,r)$ contains a unique copy of $GR(p^n,s)$.

(XVI.8) THEOREM. Let n_1,n_2,r_1,r_2 be positive integers. Let $n = \min(n_1 n_2)$, $d = \text{g.c.d.}(r_1,r_2)$ and $m = \text{l.c.m.}(r_1,r_2)$. Then

$$GR(p^{n_1},r_1) \otimes_Z GR(p^{n_2},r_2) \simeq \oplus \sum_1^d GR(p^n,m) \qquad \text{(as rings).}$$

Proof. We first consider $GR(p^n,d) \otimes_{Z/Zp^n} GR(p^n,d)$. We know that

$$GR(p^n,d) \simeq (Z/Zp^n)[X]/(f)$$

where f is a basic monic irreducible in $(Z/Zp^n)[X]$ of degree d . Let $\overline{X}$ denote the coset $X + (f)$ and r be in $GR(p^n,d)$. It is easy to check that the mapping

$$\psi : \overline{X} \otimes r \to r\overline{X}$$

induces a ring isomorphism

$$(Z/Zp^n)[X]/(f) \otimes GR(p^n,d) \to GR(p^n,d)[X]/(f) ,$$

i.e.,

$$GR(p^n,d) \otimes GR(p^n,d) \simeq GR(p^n,d)[X]/(f) .$$

But in $GR(p^n,d)[X]$, f splits into distinct linear factors

$$f = (X - a_1)(X - a_2) \cdots (X - a_d) .$$

Thus

$$GR(p^n,d)[X]/(f) \simeq \oplus \textstyle\sum_{i=1}^{d} GR(p^n,d) \qquad \text{(as rings).}$$

We now do the general case. First note

$$GR(p^{n_1},r_1) \otimes_Z GR(p^{n_2},r_2) \simeq GR(p^n,r_1) \otimes_{Z/Zp^n} GR(p^n,r_2) .$$

Since, by the lemma, both $GR(p^n,r_1)$ and $GR(p^n,r_2)$ contain copies of $GR(p^n,d)$, we have

$$GR(p^n,r_1) \otimes_{Z/Zp^n} GR(p^n,r_2)$$

$$\simeq (GR(p^n,r_1) \otimes_R R) \otimes_{Z/Zp^n} (R \otimes_R GR(p^n,r_2))$$

$$\simeq GR(p^n,r_1) \otimes_R (R \otimes_{Z/Zp^n} R) \otimes_R GR(p^n,r_2)$$

$$\simeq GR(p^n,r_1) \otimes_R (\oplus \textstyle\sum_{i=1}^d R) \otimes_R GR(p^n,r_2) \qquad \text{(by the above)}$$

$$\simeq \oplus \textstyle\sum_{i=1}^d GR(p^n,r_1) \otimes_R GR(p^n,r_2)$$

where $R = GR(p^n,d)$.

Thus it remains only to examine

$$GR(p^n,r_1) \otimes_{GR(p^n,d)} GR(p^n,r_2) .$$

Since $m = \text{l.c.m.}(r_1,r_2)$, $GR(p^n,m)$ contains naturally $GR(p^n,r_1)$ and $GR(p^n,r_2)$ and $GR(p^n,r_1) \cap GR(p^n,r_2) = GR(p^n,d)$.

Let

$$R = \{\textstyle\sum_i a_ib_i \mid a_i \text{ is in } GR(p^n,r_1) , b_i \text{ is in } GR(p^n,r_2)\} .$$

Then R is a subring of $GR(p^n,m)$, hence by (XVI.7) a Galois ring of the form $GR(p^n,t)$ with t dividing m . Since R contains $GR(p^n,r_1)$ and $GR(p^n,r_2)$, we have that r_1 and r_2 divide t . Since $m =$ l.c.m.(r_1,r_2) , we must have $m = t$ and $R = GR(p^n,m)$.

Define

$$GR(p^n,m) \to GR(p^n,r_1) \otimes GR(p^n,r_2)$$

by

$$\sum a_i b_i \to \sum a_i \otimes b_i \ .$$

It is easy to check that this is a surjective ring morphism. Further

$$\begin{aligned} |GR(p^n,r_1) \otimes GR(p^n,r_2)| &= |GR(p^n,d)|^{r_1 r_2} \\ &= (p^n)^{dr_1 r_2} \\ &= p^{nm} \\ &= |GR(p^n,m)| \ . \end{aligned}$$

Thus, since any surjection between finite sets of the same cardinality is a bijection, the map is an isomorphism.

We conclude this section with a determination of the group of units of a Galois ring. For a ring R , we let R^* denote the group of units of R .

<u>(XVI.9)</u> <u>THEOREM</u>. (<u>Units of a Galois Ring</u>.) Let $R = GR(p^n,r)$. Then

$$R^* = G_1 \times G_2 \quad \text{(direct product of groups)}$$

where

(a) G_1 is a cyclic group of order $p^r - 1$

(b) G_2 is a group of order $p^{(n-1)r}$ such that

(1) If p is odd, or if $p = 2$ and $n \leq 2$, then G_2 is a direct product of r cyclic groups each of order p^{n-1} .

(2) If $p = 2$ and $n \geq 3$, then G_2 is a direct product of a cyclic group of order 2 , a cyclic group of order 2^{n-2} and $r - 1$ cyclic groups each of order 2^{n-1} .

Proof. (Raghavendran). The case $n = 1$ is handled in (II). The case $r = 1$ is well-known from classical number theory and may be deduced, with modifications, from the following proof. Thus, we suppose $n \geq 2$ and $r \geq 2$.

For simplicity let R denote $GR(p^n,r)$ and K denote R/pR .

From the natural surjective group morphism $\mu : R^* \to K^*$ it is clear that

$$\begin{aligned} |R^*| &= |K^*||pR| \\ &= (p^r - 1)(p^{(n-1)r}) . \end{aligned}$$

Since $(p^r - 1, p^{(n-1)r}) = 1$,

$$R^* = G_1 \times G_2$$

where $|G_1| = p^r - 1$ and $|G_2| = p^{(n-1)r}$.

The structure of G_1 is easy. Let $K^* = \langle\bar{a}\rangle$. Let a be in G_1 with $\mu a = \bar{a}$. Note that the order of a is at least $p^{r-1} = |G_1|$ which is the order of $\bar{a}$. Thus, $G_1 = \langle a\rangle$.

<u>Case I</u>. $(p = 2, n \geq 3)$. Since in K , $(0)^2 + 0 = (1)^2 + 1$, the map $K \to K$ given by $a \to a^2 + a$ is not injective. Consequently, the map is not surjective. Thus, there is a $\bar{b}$ in K such that $X^2 + X = \bar{b}$ has no solution in K . Select b in R with $\mu b = \bar{b}$.

Now let $g_1 = 1, g_2, \ldots, g_r$ be a free $Z/Z2^n$-basis for $R = GR(2^n, r)$. Observe $-1 + 2^{n-1}g_1$ is in G_2 , $(-1 + 2^{n-1}g_1)^2 = 1$, $(1 + 4b)^{2^{(n-2)}} = 1$ and that $a^{2^{(n-1)}} = 1$ for all a in G_2 .

We claim that if $m, n_1, n_2, \ldots, n_r$ are positive integers with

$$m \leq 2 \ , \quad n_1 \leq 2^{n-2} \ , \quad n_i \leq 2^{n-1} \qquad (2 \leq i \leq r)$$

and

$$(**) \qquad (-1 + 2^{n-1}g_1)^m (1 + 4b)^{n_1} \prod_{i=2}^{r} (1 + 2g_i)^{n_i} = 1$$

then

$$m = 2 \ , \quad n_1 = 2^{n-2} \ , \text{ and } \ n_i = 2^{n-1} \qquad (2 \leq i \leq r) \ .$$

To prove this we need the following lemma. The proof of the lemma is easy.

(XVI.10) LEMMA. Let p be an odd prime and let a_t, b_t, c_t be the coefficients of X^t in the expansions of $(1 + pX)^N, (1 + 2X)^N, (1 + 4X)^N$ respectively. Then,

(a) If $p^\alpha | N$ then $p^{\alpha+1} | a_1$ and $p^{\alpha+2} | a_t$ for all $t \geq 2$,

(b) If $2^\alpha | N$ then $2^{\alpha+1} | b_t$ for $t = 1,2$ and $2^{\alpha+2} | b_t$ for $t \geq 3$,

(c) If $2^\alpha | N$ then $2^{\alpha+2} | c_1$ and $2^{\alpha+3} | c_t$ for $t \geq 2$,

(d) $4 | b_t$ for all $t \geq 2$.

We now prove the assertion. Suppose that $m = 1$. Using (d) and expanding (**) we obtain

$$2(1 + \textstyle\sum_{i=2}^{r} n_i g_i + 2a) = 0$$

for some a in R . Thus $(1 + \sum_{i=2}^{r} n_i g_i + 2a)$ is in (2) and

$$1 + \textstyle\sum_{i=2}^{r} n_i \mu g_i = 0 .$$

Since $\{1 = \mu g_1, \mu g_2, \ldots, \mu g_r\}$ are $Z/Z2$-free we have a contradiction. Hence $m = 2$ and (**) reduces to

$$(***) \qquad (1 + 4b)^{n_1} \prod_{i=2}^{r} (1 + 2g_i)^{n_i} = 1 .$$

By the argument it is also evident that $n_2, \ldots, n_r$ are even.

Let α be the integer $0 \leq \alpha \leq n - 2$ such that $2^{\alpha+1}$ is the highest power of 2 which divides each of the integers $2n_1, n_2, \ldots, n_r$. It is our purpose to show that $\alpha = n - 2$. Let $n_1 = 2^\alpha m_1$, $n_i = 2^{\alpha+1} m_i$ for $i \geq 2$. It is clear that one of $\{m_1, \ldots, m_r\}$ is odd.

We now apply to (XVI.10)(b) (with $\alpha + 1$ in the place of α) and (c). Then (***) reduces to

$$2^{\alpha+2}(m_1 b + \textstyle\sum_{i=2}^{r} m_i g_i + \sum_{i=2}^{r} m_i(m_i 2^{\alpha+1} - 1)g_i^2 + 2B) = 0$$

for some B in R . If $\alpha + 2 < n$ then

$$m_1\mu b + \textstyle\sum_{i=2}^{r} m_i\mu g_i + \sum_{i=2}^{r} m_i\mu g_i^2 = 0 .$$

By the choice of b , m_1 must be even. Then one of the m_i , $2 \leq i \leq r$, must be odd and we obtain

$$\textstyle\sum_{i=2}^{r} m_i\mu g_i = 0 \quad \text{or} \quad \sum_{i=2}^{r} m_i\mu g_i = 1 = \mu g_1 ,$$

both of which are contradictions. Thus $\alpha = n - 2$. This proves the assertion concerning (**).

Let
$$H_0 = \langle -1 + 2^{n-1}g_1 \rangle , \quad |H_0| = 2$$
$$H_1 = \langle 1 + 4b \rangle , \quad |H_1| = 2^{n-2}$$
$$H_i = \langle 1 + 2g_i \rangle , \quad |H_i| = 2^{n-1}$$

for $2 \leq i \leq r$.

It is clear that $H_0,\ldots,H_r$ are cyclic of the above orders and the above assertion implies that the product of $H_1,\ldots,H_r$ is direct. By a direct count

$$|G_2| = |H_0 \times \cdots \times H_r|$$

so that the product $H_0 \times \cdots \times H_r$ exhausts the group G_2. This completes case (I).

Case II. $(p \geq 3)$. Here consider the equation

$$\prod_{i=1}^{r} (1 + pg_i)^{n_i} = 1$$

and use (a) of (XVI.10).

Case III. $(p = 2, n = 2)$. Here the square of every element of G_2 equals 1 so G_2 will be an elementary Abelian group.

The exercises in this section are formulated for Galois rings. Many could be stated for any local ring with a principal maximal ideal.

(XVI.1) EXERCISE. An $n \times m$ matrix A over a commutative ring is said to admit diagonal reduction if there exist invertible matrices P and Q with $PAQ = \text{diag}[d_1, d_2, \ldots]$ where d_i divides d_{i+1}.

(a) (Kaplansky). Show that if all 1×2, 2×1 and 2×2 matrices over a commutative ring admit diagonal reduction, then all matrices admit diagonal reduction.

(b) Show a matrix over a Galois ring admits diagonal reduction. (Hint: Use the fact that every element in a Galois ring is of the form up^t.)

(XVI.2) EXERCISE. Let $R = GR(p^n,r)$ be a Galois ring.

(a) Show every ideal of R is of the form $Rp^t = (p^t)$ for $0 \le t \le n$.

(b) Show every non-zero element in R may be written as up^t where u is a unit and $0 \le t \le n - 1$. Show that in the representation of an element as up^t the integer t is unique and u is unique modulo (p^{n-t}).

(c) Select a set of representatives T in R of the cosets $R/Rp = GF(p^r)$. The set T is called a transversal. Show that each element in R may be written uniquely as

$$t_0 + t_1p + \cdots + t_{n-1}p^{n-1}$$

with t_i in T.

(d) Show for each positive integer d there is a natural injective ring morphism $GR(p^n,r) \to GR(p^n,dr)$. Show there is a natural surjective ring morphism $GR(p^n,r) \to GR(p^{n-1},r)$ with kernel (p^{n-1}).

(e) Show every subring of $GR(p^n,r)$ is a Galois ring of the form $GR(p^n,s)$ where s divides r. Conversely, if s divides r then $GR(p^n,r)$ contains a unique copy of $GR(p^n,s)$. Thus the number of subrings of $GR(p^n,r)$ is the number of positive divisors of r.

(XVI.3) EXERCISE. Prove Theorem (XVI.4).

(XVI.4) EXERCISE. Show a Galois ring is a homomorphic image of a principal ideal domain.

(XVI.5) EXERCISE. (Janusz [68].) Let h_n denote the natural projection $h_n : GR(p^n,r) \to GR(p^{n-1},r)$. The collection of rings and maps $\{GR(p^n,r);h_n\}$ forms an inverse system. We can form the projective limit

$$D_p(r) = \text{proj lim}\{GR(p^n,r);h_n\} .$$

Show $D_p(1)$ is the ring of p-adic integers. (For a discussion of $D_p(r)$ see Janusz [68].)

(XVI.6) EXERCISE. If r and s with $r < s$ are positive integers then r! divides s! . Thus we have the natural inclusions

$$GR(p^n,1!) \subseteq GR(p^n,2!) \subseteq GR(p^n,3!) \subseteq \cdots .$$

Let $GR(p^n,\infty) = \bigcup_r GR(p^n,r!)$.

(a) Show $GR(p^n,\infty)$ is a ring and $GR(p^n,r) \subseteq GR(p^n,\infty)$ for every r .

(b) Show that every polynomial f in $(Z/Zp^n)[X]$ which is irreducible modulo p splits into distinct linear factors in $GR(p^n,\infty)$.

(XVI.7) EXERCISE. It is well-known that subspaces of a finite dimensional vector space over a field may be identified with row spaces of matrices and in turn with the standard row canonical form of matrices under row operations (see Exercise (II.12)). Let $R = GR(p^n,t)$ be a Galois ring. If a is in R and $a = up^t$ where u is a unit (see Exercise (XVI.2)(b)) define $d(a) = t$.

(a) (Fuller Canonical Form [113].) Let A be an $m \times m$ matrix over R. Show that A is the left associate (i.e., row equivalent) of a <u>unique</u> matrix $[\alpha_{ij}]$ satisfying:

(1) $d(\alpha_{ij}) \geq d(\alpha_{ii})$ for all i and j.

(2) If $\alpha_{jj} \neq 0$ then $d(\alpha_{ji}) > d(\alpha_{jj})$ for $i < j$.

(3) Every diagonal element is in the form p^t, $0 \leq t \leq n$. (Note: $p^n = 0$.)

(4) If $i \neq j$ and $d(\alpha_{ij}) \geq d(\alpha_{jj})$, then $\alpha_{ij} = 0$. If $d(\alpha_{ij}) < d(\alpha_{jj})$, then α_{ij} is unique modulo α_{jj}.

(b) Let $P(r_0,r_1,\ldots,r_n;m)$ denote the number of $m \times m$ Fuller canonical matrices having the property that p^i occurs r_i times along the diagonal. Let $P(m)$ denote the total number of canonical matrices. Show

$$P(m) = \sum_{r_0+\cdots+r_n=m} P(r_0,r_1,\ldots,r_n;m) .$$

(c) Let $P_t^{(m)} = P(r_0,r_1,\ldots,r_t - 1,\ldots,r_n;m)$ where $P_t^{(m)} = 0$ if $r_t = 0$. Let $M_t = \prod_{i=0}^{t} q^{(t-i)r_i}$ where $M_0 = 1$ and $N_t = \prod_{i=t+1}^{n} q^{(i-t-1)r_i}$ where $N_n = N_{n-1} = 1$ and $q = |R/Rp|$. Show

$$P(r_0,r_1,\ldots,r_n;m + 1) = \textstyle\sum_{t=0}^{n} M_tN_tP_t^{(m)} .$$

(If $n = 1$ (the finite field case) then the above formula gives

$$P(r_0,r_1;m+1) = P(r_0 - 1,r_1;m) + q^{r_0}P(r_0,r_1 - 1;m)$$

which gives the number of $(m+1) \times (m+1)$ row canonical matrices of rank r_0 and size $m+1$ in terms of the number of canonical matrices of ranks $r_0 - 1$ and r_0 and size m .)

The above formulas may be thought of as counts of submodules of a module over R . For additional results and detail see [126], [127] and [128]. In Exercise XVI.1, diagonal reduction partitions the square matrices over R into equivalence classes. The number of matrices in each class and related questions are examined in [128].

(d) Let F be a free R-module of R-dimension m . Count the subspaces of $F/(p)F$ and use this to determine the number of submodules which are free direct summands of F .

<u>(XVI.8)</u> <u>EXERCISE</u>. Let F be a free R-module of dimension n where $R = GR(p^n,r)$. An R-linear morphism $\rho : F \to F$ is called <u>regular</u> if the R-algebra $R[\rho]$ is R-free in $End_R(F)$. Show that the kernel of the substitution morphism

$$\sigma_\rho : R[X] \to R[\rho]$$

given by

$$\sigma_\rho : X \to \rho$$

is a principal ideal generated by a monic polynomial. After an examination of factorization of polynomials in polynomial rings over local rings, the above will permit the development of canonical forms under similarity for regular R-morphisms.

<u>(XVI.9)</u> <u>EXERCISE</u>. (Wilson [136], Clark [106].) Let R be a ring of characteristic p^n for p a prime. Show:

(a) If R is a local ring, then R is a Galois ring if and only if $\mathrm{Rad}(R) = Rp$.

(b) If R is a ring, then R is a direct sum of matrix rings over Galois rings if and only if $\mathrm{Rad}(R) = Rp$.

<u>(XVI.10)</u> <u>EXERCISE</u>. (Wilson.) Let $Q = M_m(R)$ be the $m \times m$ matrix ring over the Galois ring $R = GR(p^n,r)$. Let M be a finitely generated left Q-module. Show

(a) M is a direct sum of cyclic left Q-modules.

(b) M is a direct sum of $p^{j\text{-th}}$ multiples of columns of Q .

(c) If M is indecomposable then M is isomorphic to a $p^{j\text{-th}}$ multiple of a column of Q .

<u>(XVI.11)</u> <u>EXERCISE</u>. Let $Q_1 = M_{n_1}(R_1)$ and $Q_2 = M_{n_2}(R_2)$ where $R_1 = GR(p^m,r)$ and $R_2 = GR(p^n,s)$. Show

$$Q_1 \otimes_Z Q_2 \simeq \oplus \textstyle\sum_1^d M_{n_1n_2}(R_3)$$

where $R_3 = GR(p^v,u)$, $v = \min\{m,n\}$, $u = \mathrm{l.c.m.}\{r,s\}$, and $d = \mathrm{g.c.d.}\{r,s\}$.

(XVI.12) EXERCISE. (See (VIII.10).) Let $Q = M_m(R)$ where $R = GR(p^n,r)$. Let S be a Galois ring containing R with $[S:R] = t$.

(a) Show $R \subseteq S \subseteq Q$ if and only if t divides m.

(b) If $tq = m$ show the centralizer in Q of S is $C_Q(S) = M_q(S)$.

(XVI.13) EXERCISE. State and prove Theorem (VIII.11) for Galois rings.

(XVI.14) EXERCISE. State and prove Theorem (VIII.12) for Galois rings.

(XVI.15) EXERCISE. Develop a Galois theory for the matrix ring over a Galois ring (see Section (C) of VIII).

(XVI.16) EXERCISE. Let $R = GR(p^n,r)$ and $q = |R/Rp|$. (See (VIII.D).)

(a) Show $|GL_m(R)| = |R|^{m^2} \prod_{i=0}^{m-1} (1 - q^{i-m})$.

(b) Let

$$B_{ij} = I + \lambda E_{ij} \qquad (i \neq j)$$

and

$$D_m(u) = I + (u - 1)E_{mm}$$

where E_{ij} is a matrix unit in $M_m(R)$, λ is in R and u is in R^*. Show $GL_m(R)$ is generated by

$$\{B_{ij}(\lambda), D_m(u) \mid \lambda \in R,\ u \in R^*,\ i \neq j,\ 1 \leq i,\ j \leq m\}.$$

(c) Determine a p-Sylow subgroup of $GL_m(R)$.

REMARK. A careful examination of the normal subgroups and other subgroups of $GL_m(R)$ are given by Klingenberg [154] and Pomfret [160]. The group automorphisms of $GL_m(R)$ are given in [161] when $p \neq 2$ and $m \geq 3$.

<u>(XVI.17)</u> <u>EXERCISE</u>. Let $R = GR(p^n,r)$ be a Galois ring and F a free R-module of R-dimension m. Let M_1 and M_2 be submodules of F. Show the following are equivalent.

(a) $M_1 \simeq M_2$ (as R-modules).

(b) $F/M_1 \simeq F/M_2$ (as R-modules).

(Hint: To show (a) implies (b) use the fact that R is self-injective. To show (b) implies (a) see Exercise (XII.17).)

<u>(XVI.18)</u> <u>EXERCISE</u>. (Snapper [60].) Let $R = GR(p^n,r)$ be a Galois ring and $\mu : R[X] \to GF(p^r)[X]$ denote the canonical morphism. Let A be a regular ideal in $R[X]$, i.e., $\mu A \neq 0$. Further assume $A \cap R = 0$. Let $A_i = \{f \text{ in } R[X] \mid p^i f \text{ is in } A\}$.

(a) Show there is a chain of ideals

$$A = A_0 \subseteq A_1 \subseteq \cdots \subseteq A_n = R[X].$$

(b) Show that under $R[X] \to R[X]/A$ the above chain is mapped to a chain of ideals

$$0 = B_0 \subseteq B_1 \subseteq \cdots \subseteq B_n = R[X]/A$$

with $B_i = \{g \text{ in } R[X]/A \mid p^i g = 0\}$.

(c) Let f_i be a monic polynomial of least degree in A_i. Then $A_i = (f_i, N_i)$ where $N_i = A_i \cap (p)[X]$. Let

$h_i = \deg(f_{i-1}) - \deg(f_i)$ for $i = 1,2,\ldots,n$. Show [Snapper [60], Statement 8.1] that the R-module $R[X]/A$ is a direct sum of $h = h_1 + \cdots + h_n$ cyclic R-modules which are generated by the images modulo A of the polynomials $f_i, f_iX, \ldots, f_iX^{h_i-1}$ for $i = 1,2,\ldots,n$.

(d) Show $A = (f, pf_1, p^2f_2, \ldots, p^{n-1}f_{n-1})$. (Hint: Use $N_i = pA_{i+1}$. Then $A = (f,N) = (f,pA_1) = (f,p(f_1,N_1)) = (f,pf_1,pN_1) = \cdots$.)

(e) Show $R[X]/A$ is R-free if and only if $A = (f)$.

(f) Let N be a nilpotent ideal in $R[X]$. Let t be the largest exponent of p such that each polynomial f of N can be written as $f = p^tg$ with g in $R[X]$. Show the set of polynomials g form a regular ideal A and $N = p^tA$.

NOTE: This exercise is useful in examining simple extensions of R .

(XVI.19) EXERCISE. Let $R = GR(p^n,r)$ and $S = GR(q^m,s)$. Determine $\mathrm{Hom}_Z(R,S)$ for various choices of primes p and q and positive integers r,s,m and n . (See remark after (XVI.3) together with Exercise (VII.2).)

XVII. LOCAL COMMUTATIVE RINGS

In this section we witness the first use of Galois rings in the structure theory of finite rings - a local commutative ring is characterized as a homomorphic image of a polynomial ring over a Galois ring. The presence of Galois rings in finite rings will be shown to be an important tool in structural questions.

The latter portions of this section concern the classification of local principal ideal rings. These rings occur naturally in two settings: First as the coordinatizing rings of Pappian Hjelmslev planes and second in algebraic number theory as quotient rings of rings of algebraic integers in finite extensions of the rationals.

With this section and the next (where we examine the units of a commutative local ring) we conclude our discussion of the structure of commutative finite rings. By (VI) every finite commutative ring decomposes uniquely as a direct sum of local rings. Thus we are left with characterizing local commutative rings. We show below that local rings are obtained naturally as the homomorphic image of a polynomial ring over a Galois ring. Thus, the theory of local rings reduces to a determination of the primary ideals Q in $GR(p^n,r)[X_1,\dots,X_t]$ and the quotient rings $GR(p^n,r)[X_1,\dots,X_t]/Q$.

(XVII.1) THEOREM. (Structure Theorem for Local Rings.) Let R be a local commutative ring of characteristic p^n with maximal ideal m and residue field k . Let $[k:Z/Zp] = r$ and $\{u_1,\dots,u_t\}$ be a minimal R-generating set of m . Then there exists a subring T of R such that:

(a) $T \simeq GR(p^n,r)$, T is unique (absolutely!) and T is the largest Galois extension of Z/Zp^n in R .

(b) R is the ring homomorphic image of $T[X_1,\dots,X_t]$.

The Galois ring T is called the coefficient ring of R .

Proof. Let $\bar{a}$ be the generator of the group of units of k and let $\bar{f} = Irr(\bar{a},Z/Zp)$ in $(Z/Zp)[X]$. Let f be a monic pre-image of $\bar{f}$ in $(Z/Zp^n)[X]$. By (XV.1) R contains an element a with $\mu a = \bar{a}$ and $f(a) = 0$. Take $T = (Z/Zp^n)[a] \simeq (Z/Zp^n)[X]/(f)$. Then T is a Galois ring, unique by (XV.7) and clearly the maximum Galois extension of Z/Zp^n in R .

It now suffices to show that $R = T[u_1,\dots,u_t]$.

Obviously $T[u_1,\ldots,u_t]$ is a subset of R. Let c be in R. Since T maps surjectively on k there is a b in T with $c \equiv b \bmod m$. If β is the nilpotency of m construct $\{c_j\}_{j=1}^{\beta-1}$ such that

$$c \equiv c_j \bmod m^{j+1}$$

and

$$c_j \text{ is in } T[u_1,\ldots,u_t] .$$

This is done by letting $c_0 = b$ and if $j \geq 1$

$$c_j = c - \sum d_i w_i$$

where w_i is a power product of $\{u_1,\ldots,u_t\}$. There is a b_i in T with $b_i \equiv d_i \bmod m$. Thus

$$c - c_j \equiv \sum b_i w_i \bmod m^{j+2} .$$

Then set $c_{j+1} = c_j + \sum b_i w_i$. Since $m^\beta = 0$, $c = c_{\beta-1}$ and c is in $T[u_1,\ldots,u_n]$.

We thus see that if R is a local ring then $R \simeq T[X_1,\ldots,X_t]/Q$ where T is a Galois ring and Q is a primary ideal with $Q \cap T = 0$. Further the radical of Q is $(p,X_1,\ldots,X_t)$. Noting that T is the homomorphic image of $(Z/Zp^n)[X]$ we have the following corollary.

(XVII.2) COROLLARY. Let R be a local commutative ring with characteristic p^n and maximal ideal m. If $\dim(m/m^2) = t$ then R is the ring homomorphic image of

$$(Z/Zp^n)[X_1,\ldots,X_t,X_{t+1}] .$$

The emphasis of the above result was on the maximal ideal of R . If instead we examine the units R^* of R a second semi-trivial structure theorem is possible.

<u>(XVII.3)</u> <u>THEOREM</u>. Let R be a local commutative ring of characteristic p^n . If $\{a_1,\ldots,a_s\}$ are the generators of the group of units of R then R is the ring homomorphic image of $(Z/Zp^n)[X_1,\ldots,X_s]$.

<u>Proof</u>. Consider the subring $(Z/Zp^n)[a_1,\ldots,a_s]$ of R . Clearly this subring contains each unit of R . Let a be in the maximal ideal. For a unit b , $a - b = c$ is a unit. Thus $a = b + c$ is in $(Z/Zp^n)[a_1,\ldots,a_s]$.

We now use (XVII.1) in a classification of local principal ideal rings. Such a ring is called a <u>chain ring</u>.

Let R denote a local principal ideal ring with characteristic p^n , maximal ideal m and residue field k where $r = [k:Z/Zp]$. Then R contains as a coefficient ring the Galois ring $T = GR(p^n,r)$.

Let Θ be in $m - m^2$. Then $m = R\Theta$ and every ideal of R is of the form $m^i = R\Theta^i$. Let β denote the nilpotency of m ; i.e., the least positive integer such that $m^\beta = R\Theta^\beta = 0$. If a is in R then $a = u\Theta^i$ where u is a unit, i is unique and u is unique modulo $R\Theta^{\beta-i}$.

(XVII.4) LEMMA. Suppose we have the above setting. Then there exist positive integers s and t such that

(a) $R = T \oplus T\Theta \oplus \cdots \oplus T\Theta^{s-1}$ (as a T-module)

(b) $\Theta^s = p(a_{s-1}\Theta^{s-1} + \cdots + a_1\Theta + a_0)$ where a_i are in T and a_0 is a unit

(c) As T-modules

$$T\Theta^i \simeq T \ , \quad 1 \le i \le t - 1 \ ,$$

and

$$T\Theta^i \simeq Tp \ , \quad t \le i \le s - 1$$

(d) $\beta = (n - 1)s + t$, $1 \le t \le s$. If $n = 1$, $s = t = \beta$.

Proof. Since p is in the maximal ideal there is an integer s such that $p = v\Theta^s$ where v is a unit. Let $u = v^{-1}$. Then $up = \Theta^s$. Then $\Theta^{s(n-1)} \neq 0$ and $\Theta^{sn} = 0$; that is, there is a t , $1 \le t \le s$, with $\beta = s(n - 1) + t$. If $n = 1$, $p = 0$ in R and we take s to be β . By the proof of (XVII.1) we have

$$\begin{aligned} R &= T[\Theta] \\ &= T + T\Theta + T\Theta^2 + \cdots + T\Theta^{\beta-1} \\ &= (T + T\Theta + \cdots + T\Theta^{s-1}) + Rp \\ &= T + T\Theta + \cdots + T\Theta^{s-1} \end{aligned}$$

by Nakayama's Lemma. Thus, since Θ is nilpotent, u is of the form $a_0 + a_1\Theta + \cdots + a_{s-1}\Theta^{s-1}$ where a_0 is a unit. This proves (b). It remains to show (a) and (c). Define $\phi_j : T \to T\Theta^j$ by $\phi_j : x \to x\Theta^j$ and observe that $p^{n-1}\Theta^j = 0$ if and only if $j \geq t$. Then, if $j < t$ the kernel of ϕ_j is 0 and $T \simeq T\Theta^j$. While if $t \leq j < s - 1$ then the kernel of ϕ_j is Tp^{n-1} and $Tp \simeq T\Theta^j$.

To show that the sum is direct it suffices to show that

$$|R| = \prod_{i=0}^{s-1} |T\Theta^i| .$$

Since $T = GR(p^n,r)$ and thus $|T| = p^{nr}$ we have $|T\Theta^i| = p^{nr}$, $0 \leq i < t$. Further, $|Tp| = p^{(n-1)r}$ so $|T\Theta^i| = p^{(n-1)r}$ for $t \leq i < s$. Thus we need to show

$$|R| = p^{nrt}p^{(n-1)r(s-t)} = p^{r\beta} .$$

For this note that $R\Theta^i/R\Theta^{i+1}$ is a $k = R/R\Theta$-vector space of dimension 1 for $0 \leq i < \beta$. Thus $|R| = (p^r)^\beta$ since $p^r = |k|$.

For the Galois ring T, the polynomial

$$g(X) = X^s + p(a_{s-1}X^{s-1} + \cdots + a_1X + a_0)$$

in $T[X]$ where a_0 is a unit is called an Eisenstein polynomial over T. The ring $T[X]/(g)$ is called an Eisenstein extension of T. Collecting the above results we have the following characterization of chain rings.

(XVII.5) THEOREM. (Characterization of Chain Rings.) Let R be a local principal ideal ring. Suppose the maximal ideal m of R has nilpotency β, the characteristic of R is p^n and $r = [R/m:Z/Zp]$. Then there exist integers t and s such that

$$R \simeq GR(p^n,r)[X]/(g(X),p^{n-1}X^t)$$

where $t = \beta - (n - 1)s > 0$ and $g(X)$ is an Eisenstein polynomial of degree s over $GR(p^n,r)$. Conversely, any such quotient ring is a local principal ideal ring.

Since the Eisenstein polynomial $g(X)$ has the form

$$\begin{aligned} g(X) &= X^s + p(a_{s-1}X^{s-1} + \cdots + a_0) \\ &= X^s + pf(X) \end{aligned}$$

we see that the structure of R depends on the bothersome polynomial f. Observe that f is not an invariant of the ring but depends on the choice of the generator Θ of the maximal ideal m of R. We now show that if $(p,s) = 1$ then Θ may be chosen so that $g(X) = X^s + p$.

We know by the proof of (XVII.4) that $p = v\theta^s$ for a unit v in R. The polynomial $X^s - \mu v$ has a simple zero $\bar{\rho}$ in R/m since $(s,p) = 1$. Thus, by (XV.1), $X^s - v$ has a zero ρ in R. Then $p = (\rho\theta)^s$. Further, since v is a unit then ρ is a unit and $(\rho\theta) = (\theta)$. Thus take $\theta_1 = \rho\theta$.

(XVII.6) COROLLARY. Suppose we have the setting of the above theorem. If $(p,s) = 1$, then we have

$$R \simeq GR(p^n,r)[X]/(X^s + p, p^{n-1}X^t) .$$

If $(p,s) = 1$ then R is called a pure chain ring. Relatively little is known concerning chain rings where p divides s and even a partial classification would be of interest.

Chain rings occur naturally in two different settings: In algebraic number theory, as quotient rings of rings of algebraic integers in finite extensions of the rationals; in geometry, as coordinatizing of Pappian Hjelmslev planes.

If we borrow the terminology of algebraic number theory, then the integer s where $(p) = (\theta)^s$ is called the ramification index of R. The ring R is called wildly ramified or tamely ramified according as p divides s or $(p,s) = 1$.

Our next purpose is a deeper examination of pure chain rings, i.e., the tamely ramified setting.

If R is a chain ring, the integers p,n,r,s, and t where

p^n is the characteristic of R

$p^r = |R/m|$

s is the greatest integer less than or equal to the nilpotency β of m such that p is in m^s

$\beta = (n - 1)s + t \;,\quad 1 \leq t \leq s$

are called the <u>invariants</u> of R. For a given set of invariants we will attempt to determine the number of isomorphism classes of pure chain rings.

If $n = 1$, then the Galois ring in (XVII.5) is $GF(p^r)$ and $R = GF(p^r)[X]/(X^s)$. This case is trivial.

———— o o o ————

For the remaining portion of this section we assume that R is a chain ring with invariants p,n,r,s, and t where $T = GR(p^n,r)$, $n > 1$, $m = R\Theta$ is the maximal ideal of R, and $k = R/R\Theta = GF(p^r)$.

———— o o o ————

<u>(XVII.7)</u> <u>LEMMA</u>. An element Π in R is a zero of an Eisenstein polynomial over T of degree s if and only if $m = (\Pi)$.

<u>Proof</u>. Let Π be a zero of $X^s + p(b_{s-1}X^{s-1} + \cdots + b_0)$ where b_0 is a unit. Then, since $p^n = 0$,

$$\Pi^{sn} = [\Pi^s + p(\text{—})]^n = 0$$

and Π is nilpotent. Thus $w = b_{s-1}\Pi^{s-1} + \cdots + b_0$ is a unit and $\Pi^s = pw$. Also, since Π is nilpotent, $\Pi = \Theta^i u$ for some unit u . By the proof of (XVII.4) $\Theta^s = pv$ for some unit v . Thus

$$pw = \Pi^s = \Theta^{is}u^s = p^i v^i u^s$$

and $p = ap^i$ for some unit a . Since $n > 1$, $i = 1$ and $\Pi = \Theta u$; i.e., $(\Pi) = (\Theta)$. The converse is immediate.

<u>(XVII.8)</u> <u>LEMMA</u>. Let S be a chain ring with invariants p,n,r,s,t with $n > 1$. Let Π generate the maximal ideal of S and let Π satisfy the Eisenstein polynomial g(X) over T of degree s . Then $R \simeq S$ if and only if $\sigma g(X)$ has a zero in R for some σ in the Galois group $G_{Z/Zp^n}(T)$ of T .

<u>Proof</u>. Suppose σg has a zero in R . Since σg is an Eisenstein polynomial of degree s we may assume by (XVII.7) above that the zero of σg is Θ . Then

$$\sum s_i\Pi^i \to \sum \sigma(s_i)\Theta^i$$

is a ring isomorphism $S \to R$. The converse is clear.

<u>(XVII.9)</u> <u>LEMMA</u>. Let R and S be pure chain rings with the same invariants and having Eisenstein polynomials $X^s - pu$ and $X^s - pv$, respectively, where v and u are units in T . Then $R \simeq S$ (as rings) if and only if

$$X^s - u^{p^i}v^{-1}$$

has a zero modulo Tp for some $0 \le i < r$.

<u>Proof</u>. If $R \simeq S$ then by (XVII.8) there is an automorphism σ of T such that $X^s - p\sigma(u)$ has a zero Π in R . But $R = T[\Theta]$ where $\Theta^s = pv$, so by (XVII.7) $(\Theta) = (\Pi)$ and $\Pi = \Theta w$ for some unit $w = a_0 + a_1\Theta + \cdots + a_{s-1}\Theta^{s-1}$ (a_0 unit) in R . Hence

$$p\sigma(u) = \Pi^s = \Theta^s w^s = pvw^s .$$

Then $\sigma(u) \equiv vw^s \pmod{Rp^{n-1}}$ and $\sigma(u) \equiv va_0^s \pmod{Tp}$. Thus a_0 is a zero of $X^s - \sigma(u)v^{-1} \pmod{Tp}$. Since σ induces (see (XV.2)) an automorphism of T/Tp , $\sigma(u) \equiv u^{p^i} \pmod{Tp}$ for some i , $0 \le i < r$. This completes the first part.

Conversely, suppose $X^s - u^{p^i}v^{-1}$ has a zero modulo Tp . Then, by (XV.11),

$$\sigma(u) \equiv u^{p^i} \pmod{Tp}$$

for some automorphism σ of T . Thus $X^s - \sigma(u)v^{-1}$ has a zero modulo Tp . Since $(s,p) = 1$; i.e., since S is pure, the zero is simple and by (XV.1) may be lifted to a zero w in S . Then $w^s = \sigma(u)v^{-1}$. Set $\Pi = w\Theta$, then Π satisfies $X^s - p\sigma(u)$ in R and by (XVII.8), $S \simeq R$.

Let p be a prime and c be a positive integer such that $(c,p) = 1$. If $c = 1$, let $\tau(c) = 1$. If $c \neq 1$, then let $\tau(c)$ denote the order of p in the group of units $(Z/Zc)^*$ of Z/Zc . Since $|(Z/Zc)^*| = \phi(c)$ where ϕ is the Euler phi-function, we have that $\tau(c)$ divides $\phi(c)$.

(XVII.10) THEOREM. Let $N = N(p,n,r,s,t)$ denote the number of isomorphism classes of pure chain rings having invariants p,n,r,s,t . Then, if $n > 1$,

$$N = \sum_{c|d} \frac{\phi(c)}{\tau(c)}$$

where $d = (s,p^r - 1)$.

Proof. (Clark and Liang.) By (XVII.9) we see that the proof of this result depends more on solutions of polynomials over finite fields than on the chain ring. Indeed, if k is the residue field of the pure chain ring and $k^* = k - \{0\}$, then for u,v in k^* define $u \sim v$ if

$$X^s - u^{p^i}v^{-1}$$

has a solution in k for some i , $0 \le i < r$. It then suffices by (XVII.9) to show that N counts the number of $\sim$-equivalence classes in k^* .

Since k^* is cyclic, let $k^* = \langle a \rangle$. Note $|k^*| = p^r - 1$. Clearly an element in k^* has an s-th root in k^* if and only if it lies in the group $A = \langle a^d \rangle$ where $d = (s, p^r - 1)$.

Thus $u \sim v$ if and only if $u^{p^i} A = vA$ for some j , $0 \le j < r$.

Since k^* is cyclic (multiplicative) of order $q = p^r - 1$, we simplify the notation by replacing k^* by the additive cyclic group Z/Zq . Then the above relation becomes $j \sim i$ if and only if

$$j \equiv p^h i \pmod{d}$$

for some h , $0 \le h < r$. Let $[i]$ denote the $\sim$-equivalence class of i and $(d) = d(Z/Zq)$. Then, as sets

$$[i] = i + (d) \cup ip + (d) \cup \cdots \cup ip^{r-1} + (d) .$$

Let $(i,d) = e$, $c = d/e$ and $h = \tau(c)$. Then

$$[i] = i + (d) \cup ip + (d) \cup \cdots \cup ip^{h-1} + (d) .$$

It is easy to see that the second union is disjoint.

Since (d) has q/d elements, [i] has $hq/d = \tau(c)q/d$ elements. Further, if j is in the class [i], $(j,d) = e$ if and only if $(i,d) = e$. Thus, if $E(e)$ denotes the set of j in Z/Zq with $(j,d) = e$ then $E(e)$ is a union of classes each having $\tau(c)q/d$ elements. But $E(e)$ contains $\phi(c)q/d$ elements. Thus $E(e)$ is a union of $\phi(c)/\tau(c)$ $\sim$-equivalence classes. As e runs through the divisors of d so also does $c = d/e$. Thus

$$N = \sum_{c|d} \frac{\phi(c)}{\tau(c)} .$$

By the exercises in (I) $\sum_{c|d} \phi(c) = d$, thus $\sum_{c|d} \frac{\phi(c)}{\tau(c)} \le d$ (equality holds if and only if $\tau(c) = 1$ for all c; i.e., if and only if d divides $p - 1$).

The number $\sum_{c|d} \frac{\phi(c)}{\tau(c)}$ arises in factoring $X^d - 1$ over Z/Zp. The number $\frac{\phi(c)}{\tau(c)}$ is the number of irreducible factors of the cyclotomic polynomial Φ_c (see (II)) over Z/Zp (see Redei, Algebra, pp. 508-513).

The numbers $\frac{\phi(c)}{\tau(c)}$ also arise (not independently) in counting the number of factors of a prime p in cyclotomic fields in algebraic number theory (see Borevich and Shafarevich, Number Theory, pp. 325-328).

(XVII.1) EXERCISE. Let R be a local ring with maximal ideal m and residue field $k = R/m$. Show that a set $\{u_1,\dots,u_n\}$ is a minimal R-generating set for the ideal m if and only if their images $\{\bar{u}_1,\dots,\bar{u}_n\}$ form a k-basis for m/m^2.

(XVII.2) EXERCISE. (Clark and Liang.) Show that σ in (XVII.8) cannot always be taken to be the identity automorphism. (Hint: Let $T = GR(3^2,2) = (Z/3^2Z)[a]$ where $a^2 = -1$. Take $R = T[\Pi]$ where Π is a zero of $g(X) = X^4 - 3(a + 1)$. Take $S = T[\Theta]$ where Θ is a zero of $X^4 - 3(1 - a)$. Then, if $\sigma : T \to T$ is the automorphism given by $\sigma : a \to -a$, then σg has a zero Θ in R and, consequently, $R \simeq S$. But g has no zero in R.)

(XVII.3) EXERCISE. (Clark and Liang.) Let p divide s. Show that if either $n > 2$ or $t > 1$ and $n = 2$, then the chain rings with associated Eisenstein polynomials $X^s - p$ and $X^s - p(X + 1)$ are not isomorphic.

(XVII.4) EXERCISE. (Clark and Liang.) Show that a chain ring is determined up to isomorphism by its invariants p,n,r,s,t if and only if they satisfy one of the following:

(a) $n = 1$

(b) $(s,p^r - 1) = 1$ and $(s,p) = 1$

(c) $n = 2$, $t = 1$ and $(s,p^r - 1) = 1$.

(XVII.5) EXERCISE. Let Q denote the number of chain rings with invariants p,n,r,s,t . Show that if p divides s and if either n > 2 or n = 2 and s = t , then

$$p^r/r \leq Q \leq (1 - p^{-r})p^{r(m-s)}$$

where $m = (n - 1)s + t$.

(XVII.6) EXERCISE. (Simple Extensions.) Classify, in the sense of (XVII.1), all local commutative rings which are simple extensions of their coefficient rings. [Hint: If T is the coefficient ring of R and $R = T[\sigma]$, show that $R \simeq T[X]/Q$ where Q is a primary regular ideal whose form is given in (XIII.13).]

(XVII.7) EXERCISE. Let $R \subset S$ be local rings. Then S is called a constant field extension of R if S and R have the same residue field. Show that a local commutative ring is a constant field extension of its coefficient ring.

(XVII.8) EXERCISE. (Schneider [133].) Let T denote the set of isomorphism classes of pure chain rings having invariants p,n,r,s,t . Show there is a bijection between T and the orbits of the Galois group of the Galois ring acting on the units of the Galois ring module s-th powers.

<u>(XVII.9)</u> <u>EXERCISE</u>. (Schneider [133].) (Continuation of XVII.8.) Show there is a bijection between T and the orbits of the automorphism group of the field k acting on the units of k modulo s-th powers where k is the residue field of the Galois ring.

<u>(XVII.10)</u> <u>EXERCISE</u>. What can be said of the wildly ramified chain rings? (Remark: See Snapper [60], p. 64.)

(For further results on chain rings, see McLean [56], Snapper [60], Clark and Drake [107], Clark and Liang [108], Fisher [114] and Schneider [133].)

XVIII. THE GROUP OF UNITS OF A COMMUTATIVE LOCAL RING

This section concerns the structure of the group of units R^* of a commutative local ring R. If m denotes the maximal ideal of R, it is quickly shown that the structure of R^* depends on the multiplicative Abelian group $1 + m = \{1 + x \mid x \text{ in } m\}$. Best results are obtained when $1 + m$ has a small number of generators. The analysis of $1 + m$ is also shown to be related to that of the additive Abelian group m. We complete our discussion of finite commutative rings in this section. The remainder of this monograph will examine further aspects of non-commutative finite rings.

The group of units of a commutative ring has for several reasons interested numerous researchers. First, there is an old question of Fuch's, "Which Abelian groups are the group of units of a commutative ring?" Second, a principal tool in the study of structures over finite fields is the fact that the group of units of a finite field is well-known - it is a cyclic group. For generalizations a a knowledge of the structure of the group of units is often essential.

Third is the observation that the structure of the units is closely related to the structure of the ring. Indeed, one might hope that the theory of finite Abelian groups and the knowledge of local commutative rings could be so blended as to provide a description of both the units and the ring. However, (XVI.7) should indicate that the problem is non-trivial even for well-conditioned rings. This section touches only some basic ideas of the theory.

Let R^* denote the group of units of a commutative ring R. We immediately reduce our considerations to the case where R is local by (VI.2) and the following easy result.

<u>(XVIII.1)</u> <u>THEOREM</u>. Let $R = R_1 \oplus \cdots \oplus R_t$ where R_i, $1 \le i \le t$, are local rings. Then R^* is the direct product of groups

$$R^* = R_1^* \times \cdots \times R_t^* .$$

——— o o o ———

<u>NOTATION</u>. We assume throughout this section that R is a local commutative ring such that

m is the maximal ideal of R

$k = R/m$ is the residue field of R and $|k| = p^t$

$p^n = \chi(R)$ is the characteristic of R

R^* is the group of units of R

$$N = \{x \text{ in } m \mid px = 0\}$$

$$Q = \{x \text{ in } N \mid x^p = 0\} .$$

o o o

For a set S in R let $a + S = \{a + s \mid s \text{ in } S\}$. Observe we have the natural chain of groups

$$1 + Q \leq 1 + N \leq 1 + m \leq R^* .$$

(XVIII.2) THEOREM. $R^* \simeq (1 + m) \times k^*$.

Proof. Observe the natural ring morphism $R \to k$ induces an exact sequence of groups

$$1 \to 1 + m \to R^* \to k^* \to 1 .$$

But k^* is cyclic of order $p^t - 1$ and $1 + m$ is a p-group. Thus, the sequence splits and

$$R^* \simeq (1 + m) \times k^* .$$

Thus the analysis of R^* is reduced to a study of k^* (which is known) and an examination of the Abelian p-group $1 + m$ (which is unknown). The group $1 + m$ is called the *one group* of R .

Observe m , N and Q are additive Abelian p-groups and $1 + m$, $1 + N$ and $1 + Q$ are multiplicative p-groups. We let $\langle a \rangle$ denote the subgroup generated by a , " $\oplus$ " denote the direct sum of additive groups and " $\times$ " denote the direct product of multiplicative groups. There may occasionally be confusion as we move between the additive groups and additive notation and multiplicative groups and multiplicative notation. The general approach is to examine the relations between the structure of m and the structure of $1 + m$.

Recall that a finite Abelian p-group G is a direct sum (product) of cyclic p-groups; and, in any two decompositions of G we have the same number of summands (factors) of each order. Indeed, written additively, this is no more than the theory of Z/Zp^n-modules. The number of cyclic summands (factors) of G is called the rank of G , denoted $rk(G)$.

Let

$$x = rk(1 + m)$$

$$y = rk(1 + N)$$

$$z = rk(m) \ .$$

Observe m and N have the same rank. Note that a p-group has rank q if and only if it admits exactly p^q solutions to $X^p = 1$, $(pX = 0$ for additive groups). Our first purpose is to show that if x , y or z is sufficiently small then $x = y = z$. Precisely, if either x , y or z is less than $p - 1$ then $x = y = z$. This has proven useful in characterizing the rings whose groups of units are generated by few elements. The proof of this result requires two technical lemmas.

If w is in N, $(1 + w)^p = 1 + w^p$. Thus $(1 + w)^p = 1$ if and only if w is in Q. Thus Q has p^y elements. If $|N| = p^z$ then $N = Q$ if and only if $y = z$.

Finally, if a is in m, let $n(a)$ denote the least positive integer v with $a^v = 0$; i.e., $n(a)$ is the <u>nilpotency</u> of a.

<u>(XVIII.3)</u> <u>LEMMA</u>.

(a) If a^m is in N and $a^w \neq 0$ for some $w \geq m$, then

$$\langle a^m\rangle + \cdots + \langle a^w\rangle \leq N$$

is a direct sum and $w - m + 1 \leq z$.

(b) If $z < p$ then $y = z$.

(c) If a is in N and $1 + a$ has order p^w, then $p^{w-1} < n(a) \leq p^w$.

(d) If p^α is the maximal order of elements in $1 + N$, then $p^{\alpha-1} \leq z$.

<u>Proof</u>. Suppose $r_m a^m + \cdots + r_w a^w = 0$ for integers $r_m, \ldots, r_w$. Let $q = n(a) - 1 \geq w$. In succession multiply by $a^{q-m}, a^{q-m-1}, \ldots, a^{q-w}$. One obtains the fact that p divides $r_m, r_{m+1}, \ldots, r_w$. Thus $r_i a^i = 0$ since a is in N for $m \leq i \leq w$. This gives part (a). For a in N, $n(a) \leq z + 1$ by (a). Thus if $z < p$ then $n(a) \leq p$ and so $N = Q$ and we have (b).

For (c) suppose that v is a power of p. Then if a is in N, $(1 + a)^v = 1 + a^v$. Thus (c) follows. Finally select a in N such that $1 + a$ has order p^α. Then, by (c), $p^{\alpha-1} < n(a)$ and, by (a), $n(a) \leq z + 1$. Thus $p^{\alpha-1} \leq z$.

For a real number r let $[\![r]\!]$ denote the greatest integer less than or equal to r.

<u>(XVIII.4)</u> <u>LEMMA</u>.

(a) If a is in N then $n(a) \leq [\![y/(p - 1)]\!] + y + 1$.

(b) If $y < p - 1$ then $y = z$.

(c) If $w = [\![(z - 1)/y]\!]$ then $p^w \leq z$.

(d) If $z < p - 1$ then $x = y = z$.

<u>Proof</u>. Let a be in N. Observe that if $n(a) \leq rp$ for an integer r, then $n(a) \leq r + y$. Indeed, by (XVIII.3) if $n(a) > r + y$ then $\langle a^r \rangle + \cdots + \langle a^{r+y} \rangle \leq Q$ is a direct sum. Then $|Q| > p^y$. Now suppose that

$$n(a) > [\![y/(p - 1)]\!] + y + 1$$

and let $\beta = n(a)$. Then if $r = [\![(\beta - 1)/p]\!] + 1$, $\beta \leq [\![(\beta - 1)/p]\!] + 1 + y$. But, then,

$$n(a) \leq 1 + y + [\![y/(p - 1)]\!]$$

- a contradiction. Thus we have (a). If $y < p - 1$ then for a in

N by (a) we have $n(a) \leq y + 1 < p$. Thus $Q = N$ and $y = z$. This gives (b). If $\alpha y \geq z$, then $\alpha \geq z/y$ and hence $\alpha \geq [\![(z + y - 1)/y]\!]$. Thus to obtain (c) apply (d) of (XVIII.3).

It remains to show (d). By (b) of (XVIII.3) it suffices to show that $x = y$. Suppose that $x > y$. Then for some w in m , $pw \neq 0$ and $(1 + w)^p = 1$. Hence

$$1 = (1 + w)^p = 1 + pw + \cdots + w^p$$

and

$$w^p = - pw - \sum_{i=2}^{p-1} t_i w^i$$

where t_i is an appropriate binomial coefficient which is divisible by p .

Let $t_i = ps_i$ and $c = -1 - \sum s_i w^{i-1}$. Then $w^p = pwc$. If $\beta = n(w)$ then $0 = w^\beta = cpw^{\beta-p+1}$. Since c is a unit, $w^{\beta-p+1}$ is in N . But $\beta - p + 1 \leq \beta - 1$ and $w^{\beta-1} \neq 0$. Now apply (a) of (XVIII.3) and

$$z \geq (\beta - 1) - (\beta - p + 1) + 1 = p - 1$$

- a contradiction.

<u>(XVIII.5)</u> <u>THEOREM</u>. (Pearson and Schneider). Let $m \neq 0$ and $1 \leq w < p - 1$. Then the following are equivalent:

(a) rank $(1 + m) = w$

(b) rank $(1 + N) = w$

(c) rank $(m) = w$.

Proof. Let $1 \leq w < p - 1$. If $z = w$ apply (XVIII.4) (d). If $y = w$ then (b) of (XVIII.4) implies $z = y$. Then use again (XVIII.4) (d). If $x = w$, then $y \leq x = w < p - 1$. Then, as above, $y = z$ and apply (XVIII.4) (d).

We now turn to an old problem, "Classify all local rings having a cyclic group of units." Perhaps the first determination of these rings was due to Gilmer [153]. Periodically new proofs occur in the literature.

(XVIII.6) THEOREM.

(a) There is a natural group isomorphism

$$m^i/m^{i+1} \to (1 + m^i)/(1 + m^{i+1})$$

for each integer $i \geq 1$ where

$$x + m^{i+1} \to (1 + x)(1 + m^{i+1}) .$$

(Note m^i/m^{i+1} is an additive group and $(1 + m^i)/(1 + m^{i+1})$ is a multiplicative group.)

(b) m^i/m^{i+1} (and consequently, $(1 + m^i)/(1 + m^{i+1})$) is an elementary Abelian p-group.

Proof. Define a mapping

$$\eta : m^i \to (1 + m^i)/(1 + m^{i+1})$$

by

$$\eta : x \to (1 + x)(1 + m^{i+1})$$

for x in m^i . Since $1 + m^i$ is a multiplicative group for each x and y in m^i there is a z in m^i such that $(1 + x + y)(1 + z) = 1$. Then

$$(1 + x)(1 + y) = (1 + x + y)(1 + (1 + z)xy) .$$

Thus since $1 + (1 + z)xy$ is in $1 + m^{i+1}$ we have that $\eta(x)\eta(y) = \eta(x + y)$; i.e., η is a group morphism. Then $\ker(\eta) = \{x$ in $m^i \mid 1 + x$ is in $1 + m^{i+1}\}$. That is, $\ker(\eta) = m^{i+1}$. Since η is surjective we have (a).

To show (b) note that m^i/m^{i+1} is naturally a $k = R/m$-vector space. Since k has characteristic p , $p(m^i/m^{i+1}) = 0$, thus m^i/m^{i+1} is elementary Abelian.

Let $\sigma(u)$ denote the cyclic group of order u .

<u>(XVIII.7)</u> <u>COROLLARY</u>. Let R be a local commutative ring with maximal ideal m such that $m^2 = 0$. Then

$$R^* \simeq \left(\prod_{i=1}^{qt} \sigma(p)\right) \times \sigma(|k| - 1)$$

where $q = \dim_k(m/m^2)$ and $|k| = p^t$.

<u>Proof</u>. By (XVIII.2)

$$R^* \simeq (1 + m) \times k^*$$
$$\simeq (1 + m)/(1 + m^2) \times k^*$$

since $m^2 = 0$. Now apply (XVIII.6).

The direct product of two cyclic groups is a cyclic group if and only if their orders are coprime. Thus since $R^* \simeq (1 + m) \times k^*$, k^* is cyclic and $(|1 + m|, |k^*|) = 1$, then R^* is cyclic when $m \neq 0$ if and only if $1 + m$ is cyclic.

(XVIII.8) THEOREM. (Ayoub). Assume $m \neq 0$.

(a) If either m or $1 + m$ is cyclic then R is a chain ring with $R/m = Z/Zp$.

(b) If $p \geq 3$ then m is cyclic if and only if $1 + m$ is cyclic.

(c) If $p = 2$ then:

(1) If m is cyclic then $1 + m$ is cyclic if and only if $m^2 = 0$. If $m^2 \neq 0$ then $1 + m = \langle -1 \rangle \times (1 + m^2)$ where $1 + m^2$ is cyclic.

(2) If $1 + m$ is cyclic and m is not cyclic then $m \simeq (Z/2Z) \oplus (Z/2Z)$.

Proof. To show (a), since $m^i/m^{i+1} \simeq (1 + m^i)/(1 + m^{i+1})$, either hypothesis implies that m^i/m^{i+1} is cyclic. But m^i/m^{i+1} is elementary Abelian and hence $m^i/m^{i+1} \simeq Z/Zp$. It is easy then to see that R is a chain ring and $R/m \simeq Z/Zp$.

Part (b) follows from (XVIII.5).

The remainder of the proof follows from Exercise (XVIII.1).

To show (c) (1) suppose m is cyclic. If $m^2 = 0$ then $m \simeq 1 + m$ and $1 + m$ is cyclic. Suppose $m^2 \neq 0$. By Exercise (XVIII.1) $1 + m^2$ is cyclic. Observe $-1 = 1 + (-2)$ is in $1 + m$. But, if -1 is in $1 + m^2$ then 2 is in $1 + m^2$. Then $2 = 2a$ for some a in m since $m^2 = 2m$. But then $2(1 - a) = 0$ and since $1 - a$ is a unit, $2 = 0$. Thus $m^2 = 2m = 0$ - a contradiction. Hence -1 is not in $1 + m^2$ and $1 + m \simeq \langle -1 \rangle \times (1 + m^2)$.

Finally assume that $1 + m$ is cyclic and m is not cyclic. Then $m^2 \neq 0$. We claim $m^3 = 0$. Suppose that $m^3 \neq 0$. If $m^3 \neq 0$ and β denotes the nilpotency of m ; i.e., least positive integer with $m^\beta = 0$, then $\beta \geq 4$. Take x in m with $2x = 0$. Then $(1 + x)^4 = 1 + x^4$ and $1 + x^4$ is in $1 + m^4$. Since $|1 + m| = 2^{\beta-4}$ (note, $\beta \geq 4$ since $m^3 \neq 0$),

$$1 = (1 + x^4)^{2^{\beta-4}} = (1 + x)^{2^{\beta-2}} .$$

Which, since $1 + m$ is cyclic of order $2^{\beta-1}$ implies that x is in m^2 . That is, for each x in m with $2x = 0$ then x is in m^2 . Thus, since by Exercise (XVIII.1) m^2 is cyclic, this implies m has a unique subgroup of order 2 . Then m is cyclic - a contradiction. Thus $\beta = 3$ and $m \simeq (Z/Z2) \oplus (Z/Z2)$.

We have as a corollary the original result of Gilmer.

(XVIII.9) THEOREM. (Gilmer). For the local commutative ring R , R^* is cyclic if and only if $1 + m$ is cyclic. If R^* is cyclic then R is isomorphic to one of the following:

(a) $GF(p^t)$ (if $m = 0$) .

(b) Z/Zp^n (if $p \geq 3$, $n > 1$) .

(c) $Z/4Z$.

(d) $(Z/pZ)[X]/(X^2)$.

(e) $(Z/2Z)[X]/(X^3)$.

(f) $(Z/4Z)[X]/(2X,X^2 - 2)$.

Further, if m is cyclic then R is isomorphic to one of the following:

(a) $GF(p^t)$ (if $m = 0$) .

(b) Z/Zp^n $(n > 1)$.

(c) $(Z/Zp)[X]/(X^2)$.

Using (XVIII.7) the proof is a straightforward checking of cases.

(XVIII.1) EXERCISE. (Ayoub). Let R be a local ring with maximal ideal m and characteristic 2^λ . Show

(a) If $1 + m$ is cyclic and m^{i+1} is cyclic for $i \geq 2$ then m^i is cyclic.

(b) If m is cyclic and $1 + m^{i+1}$ is cyclic for $i \geq 2$ then

$1 + m^i$ is cyclic.

(XVIII.2) EXERCISE. Let R be a commutative ring and suppose $R = \oplus \sum_{i=1}^{n} R_i$ where R_i are local rings. Show that R^* is cyclic if and only if each R_i^* is cyclic and $(|R_i^*|, |R_j^*|) = 1$ for $i \neq j$.

(XVIII.3) EXERCISE. Let R and S be commutative rings and $\sigma : R \to S$ a surjective ring morphism. Show that σ induces a surjective group morphism $\sigma : R^* \to S^*$. In particular, if R^* is cyclic then S^* is cyclic.

(XVIII.4) EXERCISE. Show that $R = \left\{ \begin{bmatrix} a & b \\ 0 & c \end{bmatrix} \mid a,b,c \text{ in } Z/2Z \right\}$ is a non-commutative local ring with a cyclic group of units. (This is the only indecomposable non-commutative ring having a cyclic group of units.)

(XVIII.5) EXERCISE. (On Non-finite Rings). Show that an infinite local commutative ring cannot have a cyclic group of units. Let R be a not necessarily finite ring with descending chain condition. Show that if R^* is cyclic then R is finite. Give an example of a non-finite commutative ring without descending chain condition and having a cyclic group of units.

(XVIII.6) EXERCISE. For a local commutative ring R with maximal ideal m, let $\sigma_i : R/m^i \to R/m^{i-1}$ denote the natural morphism for $2 \leq i \leq \beta$ (β = nilpotency of m). Show that if u is a unit in R/m^{i-1} with order t and $\sigma_i(w) = u$ then w is a unit in R/m^i and the order of w is t or p^t. (p = characteristic of R/m)

(XVIII.7) EXERCISE. In the notation of (XVIII.5) show: (p is a prime)

(a) $R = (Z/pZ)[X]/(X^{p+1})$ has $z = p$ and $x = p - 1$.

(b) If $n \neq p$ and $n \neq 4$, then $R = (Z/nZ)[X]/(([1 + X]^p - 1)/X))$ has $x > p - 1 = z$.

This shows that in some sense (XVIII.5) is the best possible.

(XVIII.8) EXERCISE. Assume the setting is that of the beginning of this section. If $n(a) > z$ for some a in N then show $y = z - [\![z/p]\!]$. Further, show that in this case:

(a) If $X(R) = p$, $R \simeq (Z/Zp)[X]/(X^{z+1})$.

(b) If $X(R) = p^n$ $(n > 1)$, $R \simeq (Z/Zp^n)[X]/(pX, p^{n-1} - vX^z)$ where $1 \leq v \leq p - 1$.

(XVIII.9) EXERCISE. (Pearson and Schneider.) If $x = z = 2$ and $p = 2$, show that R is isomorphic to one of the following rings:

(a) $GF(2^2)[X]/(X^2)$.

(b) $(Z/4Z)[X]/(X^2 + X + 1)$.

(c) $(Z/8Z)[X]/(2X, X^2 + 4)$.

(d) $(Z/4Z)[X]/(2X, X^2)$.

(e) $(Z/4Z)[X]/(X^2)$.

(f) $(Z/4Z)[X]/(X^2 + 2)$.

(g) $(Z/4Z)[X]/(X^2 + 2X + 2)$.

(h) $(Z/2Z)[X,Y]/(X^2,Y^2,XY)$.

(XVIII.10) EXERCISE. (Pearson and Schneider). If $x = z = 2$ and $p \geq 3$, show that R is isomorphic to one of the following rings:

(a) $GF(p^2)[X]/(X^2)$.

(b) $(Z/Zp^n)[X]/(X^2 - a)$ where $n > 1$ and a is a non-square unit in Z/Zp^n .

(c) $(Z/Zp)[X]/(X^3)$.

(d) $(Z/Zp)[X,Y]/(X^2,Y^2,XY)$.

(e) $(Z/Zp^n)[X]/(p^rX,X^2 - wp^t)$ where $n > 1$, $1 \leq r \leq n$, $1 \leq t \leq n$, $r + t \geq n$ and w is a unit in Z/Zp^n .

REMARK: Pearson and Schneider [159] have also determined all commutative local rings with

(a) $p = 3$ and $2 = x \neq z$.

(b) $p = 2$ and $2 = x \neq z$.

This completes the classification when $x = 2$ since if $x \neq z$ and $x = 2$ then $p = 2$ or $p = 3$.

XIX. THE ROLE OF THE GALOIS RING IN THE THEORY OF FINITE RINGS

This section returns to the theory of finite non-commutative rings. The principal result concerns the existence in a ring R of a Galois subring S and the examination of R as a two-sided S-module. If R is local then S is unique up to inner-automorphism and R splits R = S ⊕ N as an (S,S)-module where N is a submodule of the maximal ideal of R. As a corollary this provides a generalization of the Wedderburn Principal Theorem to arbitrary rings. We also obtain characterizations of the ring R as (a) a homomorphic image of a skew-polynomial ring over S and (b) a homomorphic image of a subring of a matrix ring over S.

Let R be a ring. By Theorem (I.1) we assume the characteristic of R, $\chi(R) = p^{\lambda}$, is a power of a prime p.

Let S denote a local ring in R containing Z/Zp^{λ} and maximal with respect to the radical of S being pS. Since S is local, S/pS is a finite field with a cyclic group of units $(S/pS)^* = \langle\bar{\Theta}\rangle$. Let Θ be in S with $\Theta \to \bar{\Theta}$ and let T be the subring in S generated by 1 and Θ . Then $T + pS = S$. Consequently,

$$S = T + p(T + pS) = T + p^2S = \cdots = T + p^iS .$$

If $i > \lambda$ where λ is the nilpotency of p , then $T = S$. Thus S is a commutative local extension of Z/Zp^n with radical Sp , i.e., S is unramified over Z/Zp^n . Hence S is a Galois ring and is called a <u>coefficient ring</u> of R . The uniqueness of this subring will be examined when R is local, in (XIX.4).

Thus a ring R contains a Galois subring S and hence R may be regarded as a two-sided (S,S)-module. Observe that S may only be $GR(p^\lambda,1) = Z/Zp^\lambda$.

<u>(XIX.1)</u> <u>THEOREM</u>. (Rings as Images of Skew-Polynomial Rings.) (Wirt.) Let R be a ring with a coefficient ring S . Let $\mathrm{Aut}(S) = \{\sigma_1 = 1,\sigma_2,\ldots,\sigma_r\}$. Then R is the ring homomorphic image of the skew-polynomial ring

$$S[X_1,\ldots,X_t;\sigma_{i(1)},\ldots,\sigma_{i(t)}]$$

where the X_i are non-commuting indeterminates and for each i , $\sigma_{j(i)}$ is an automorphism of S satisfying

$$sX_i = X_i\sigma_{j(i)}(s)$$

for all s in S .

Proof. The mapping $1 \to 1$ determines a natural inclusion $0 \to S \to R$ as (S,S)-modules. Since a Galois ring is self-injective the sequence splits and S is a (S,S)-direct summand of R. That is, $R = S \oplus N$. By (XVI) $N = \oplus \sum_{j=1}^{r} \sum_{i=1}^{n(j)} Sb_{ij}$ where $sb_{ij} = b_{ij}\sigma_j(s)$ for all s in S. Define

$$S[X_{ij};\sigma_j] \to R$$

by $1 \to 1$ and $X_{ij} \to b_{ij}$.

A second characterization by Wilson [136] employs a right regular representation.

(XIX.2) THEOREM. (Regular Representation.) (Wilson.) Let R be a ring with a coefficient ring S. If, as a left S-module,

$$_{S}R = \oplus \sum_{i=1}^{t} Sb_i$$

then R is the ring homomorphic image of a subring of the $t \times t$ matrix ring over S.

Proof. Embed $_{S}R = \oplus \sum_{i=1}^{t} Sb_i$ as an S-submodule in the free S-module $F = \sum_{i=1}^{t} Sx_i$ (see (XVI)). Consider R as a subring of $\text{End}_S(R)$ by right regular representation, i.e. for r in R identify $r \to \sigma_r$ where $\sigma_r(x) = xr$. Define

$$T = \{\beta \text{ in } \text{End}_S(F) \mid \text{there exists an } r \text{ in } R \text{ with } \beta(x) = \sigma_r(x) \text{ for all } x \text{ in } R\}.$$

Observe if β is in $\mathrm{End}_S(F)$ <u>and</u> there exists an r in R with $\beta|_R = \sigma_r$ then r is uniquely determined by $\beta(1) = \sigma_r(1) = r$.

Further T is a ring and $T \to R$ by $\beta \to \beta(1)$ is a ring morphism. It is only necessary to show $T \to R$ is surjective. But F is an injective S-module. Thus

$$\begin{array}{ccccc} & & F & & \\ & & \uparrow{\scriptstyle \sigma_r} & & \\ 0 & \longrightarrow & R & \xrightarrow{\text{inc.}} & F \end{array}$$

can be completed to $\beta : F \to F$ with $\beta(\text{inc}) = \sigma_r$, i.e., $\beta|_R = \sigma_r$.

<u>(XIX.3)</u> <u>COROLLARY</u>. Let R be a ring with a coefficient ring S . If R is S-free of S-dimension t then R may be taken to be a subring of the $t \times t$ matrix ring over S .

Employing some of the ideas introduced in (XVI) we discuss (XIX.2) in greater detail.

Let R be a ring with a coefficient ring S . Let Q denote the injective envelope of R as an S-module. By (XVI), Q is S-free. The application of $\mathrm{Hom}_S(R,\ )$ to the exact sequence $0 \to R \to Q$ gives $\mathrm{End}_S(R) \subset \mathrm{Hom}_S(R,Q)$. By right regular representation $R \subset \mathrm{End}_S(R)$ and hence $R \subset \mathrm{Hom}_S(R,Q)$.

Next if we apply $\mathrm{Hom}_S(\ ,Q)$ to $0 \to R \to Q$ and utilize the fact that Q is injective, then

$$\mathrm{End}_S(Q) \xrightarrow{\mu} \mathrm{Hom}_S(R,Q) \longrightarrow 0$$

is exact. Then the inverse image of R under μ is a subring of $\mathrm{End}_S(Q)$. The restriction of μ to this subring is a ring morphism onto R. Since Q is free, $\mathrm{End}_S(Q)$ may be taken to be a matrix ring over S.

The above paragraph employed the injective envelope Q of R as an S-module. A similar approach can utilize the projective cover P of R as an S-module. Recall $P \simeq Q$.

Let $P \to R \to 0$ be a projective cover of R. Then there is a natural sequence of injections

$$R \to \mathrm{End}_S(R) \to \mathrm{Hom}_S(P,R)$$

where the first $R \to \mathrm{End}_S(R)$ is right regular representation and the second is given by applying $\mathrm{Hom}_S(\ ,R)$ to the exact sequence $P \to R \to 0$. Since P is projective, $\mathrm{Hom}_S(P,\ )$ applied to $P \to R \to 0$ gives an exact sequence $\mathrm{End}_S(P) \to \mathrm{Hom}_S(P,S) \to 0$ where we may take $R \subseteq \mathrm{Hom}_S(P,S)$. Again the inverse image of R is a subring of $\mathrm{End}_S(P)$ which may be taken to be a matrix ring.

If R is a local ring, we can determine the uniqueness of the coefficient ring

(XIX.4) THEOREM. Let R be a local ring with characteristic p^{λ}, maximal ideal m and residue field $R/m = GF(p^r)$. Then R has a coefficient ring S satisfying

(a) $S = GR(p^{\lambda},r)$.

(b) S is unique up to inner-automorphism of R.

(c) As an (S,S)-module,

$${}_S R_S = S \oplus N$$

where $N \subseteq m$.

Proof. Construct S as follows. The finite field $GF(p^r) \simeq (Z/Zp)[X]/(\bar{f}) \simeq (Z/Zp)[\bar{\theta}]$ for some monic irreducible $\bar{f}$ of degree r in $(Z/Zp)[X]$ and $\bar{\theta}$ in $GF(p^t)$ a zero of $\bar{f}$. Let f in $(Z/Zp^n)[X]$ be a monic pre-image of $\bar{f}$ of degree r. Hence f is irreducible and by (XV) $(Z/Zp^n)[X]/(f)$ is the Galois ring $GR(p^{\lambda},r)$. Further, $\bar{\theta}$ has a pre-image θ in T where θ satisfies f and $GR(p^{\lambda},r) \simeq (Z/Zp^{\lambda})[\theta]$. Let S denote the subring $(Z/Zp^{\lambda})[\theta]$ in R. It is clear that $S/Sp \simeq GF(p^r)$.

To show (b), recall Hall's Theorem (see, for example, [19], p. 117) concerning solvable groups; that is, if G is a solvable group of order $|G| = ab$ where $(a,b) = 1$, then G contains a subgroup of order a and any two such are conjugate under an inner-automorphism. Observe that $1 + m$ is solvable since it is a finite p-group, and $GF(p^r)^* = k^*$ is solvable since it is Abelian. Thus $R^* \simeq (1 + m) \times k^*$ is solvable.

Let H of order $|k^*|$ denote the cyclic subgroup generated by Θ in S. Let H' be any subgroup of R^* of order $|k^*|$. Since $(|k^*|, |1 + m|) = 1$, by Hall's Theorem there is an x in R^* with $xHx^{-1} = H'$. Now suppose T is a second subring of R with $T \simeq GR(p^\lambda, r)$. Noting T contains a subgroup H' of order $|k^*|$, there is an x with $xSx^{-1} \subseteq T$. But $|T| = |xSx^{-1}|$, hence $T = xSx^{-1}$.

It remains to show that, as an (S,S)-module, ${}_SR_S = {}_SS_S \oplus {}_SN_S$ where N is in the maximal ideal m. The method of proof is due to Wilson [138].

Since S is self-injective, the inclusion $0 \to S \to R$ splits as (S,S)-modules. Thus S is an (S,S)-direct summand of R.

Considering R as a left S-module

$${}_SR = Sb_1 \oplus Sb_2 \oplus \cdots \oplus Sb_t$$

where we may take $b_1 = 1$ and order the b_i so that $\text{Annih}_S(b_i) = 0$, $1 \le i \le q$, and $\text{Annih}_S(b_i) \ne 0$ for $q + 1 \le i \le t$.

Clearly b_i is in m for $q + 1 \le i \le t$.

For, $2 \le i \le t$, replace b_i by $b_i - c_i$ where c_i is in S and $\bar{c}_i = \bar{b}_i$ under $R \to R/m$. Again, call each new element " b_i ". It is easy to verify that again

$${}_SR = Sb_1 \oplus \cdots \oplus Sb_t ,$$

the annihilators are unchanged, and, in addition, $b_2,\ldots,b_t$ are in m .

Thus

$$_S m = Sp \oplus (\oplus \textstyle\sum_{i=2}^{t} Sb_i) .$$

On the other hand, m is an ideal and consequently an (S,S)-module.
Let

$$_S m_S = \oplus \textstyle\sum_{i=1}^{w} Sa_i$$

be a decomposition of m as an (S,S)-module. Since this decomposition is clearly a left decomposition, the uniqueness of left decompositions gives $w = t$ and, after reindexing, $\mathrm{Annih}_S(a_1) = (p^{\lambda-1})$ and $\mathrm{Annih}_S(a_i) = \mathrm{Annih}_S(b_i)$, $2 \le i \le t$. In particular, one checks that $\{1,a_2,\ldots,a_q\}$ are S-independent, indeed, are S-free. Therefore,

$$S1 \oplus Sa_2 \oplus \cdots \oplus Sa_q$$

is an (S,S)-submodule of R . Further, being a direct sum of injectives, this submodule is injective and hence a (S,S)-direct summand of R .

Thus,

$$_S R_S \simeq [S \oplus (\oplus \textstyle\sum_{i=2}^{q} Sa_i)] \oplus {}_S M_S$$

as an (S,S)-module.

Now express $M = \oplus \sum_{j=1}^{t-q} Sc_j$ as a direct sum of left S-modules. Then

$$_S R \simeq [S \oplus (\oplus \sum_{i=2}^{q} Sa_i)] \oplus [\oplus \sum_{j=1}^{t-q} Sc_j] .$$

Again, by uniqueness of cyclic left decompositions, $\mathrm{Annih}_S(c_j) \neq 0$ for $1 \leq j \leq t - q$. Thus the c_j's , and hence M , are in m .

Now take $_S N_S = (\oplus \sum_{i=2}^{q} Sa_i) \oplus M$.

The next result utilizes (XIX.4) to obtain a generalization of the Wedderburn Principal Theorem.

<u>(XIX.5)</u> <u>THEOREM</u>. (Wedderburn Principal Theorem.) (Clark, Wilson.) Let R be a ring of characteristic $\chi(R) = p^\lambda$. Then R contains a subring T satisfying

(a) $R = T \oplus N$ (as Abelian groups) where $\mathrm{Rad}(T) = pT$ and $N \subset \mathrm{Rad}(R)$.

(b) $R/\mathrm{Rad}(R) \simeq T/pT$.

(c) T is a direct sum matrix rings over Galois rings.

<u>Proof</u>. The orthogonal idempotents of $R/\mathrm{Rad}(R)$ may be lifted to R and provide a decomposition of R as an Abelian group

$$R = T_1 \oplus \cdots \oplus T_q \oplus \overline{N}$$

where $\overline{N} \subset \mathrm{Rad}(R)$ and T_i is a matrix ring over a local ring R_i . Let $T_i = M_{n_i}(R_i)$. By (XIX.4) R_i splits as $R_i = S_i \oplus N_i$. Take

$T = \oplus \sum_{i=1}^{q} M_{n_i}(S_i)$ and $N = (\oplus \sum_{i=1}^{q} M_{n_i}(N_i)) \oplus \overline{N}$. The remainder is straightforward.

A second application of (XIX.4) concerns local rings and the number of generators necessary in (XIX.1).

(XIX.6) THEOREM. Let R be a local ring with maximal ideal m and residue field k . Let $t = \dim_k(m/m^2)$ and S be a Galois coefficient ring of R . Then R is the ring homomorphic image of

$$S[X_1,\ldots,X_t;\sigma_1,\ldots,\sigma_t]$$

for suitable automorphisms σ_i of S .

Proof. Let ${}_SR_S = S \oplus N$ with $N \subset m$. It is easy to see that $m = pS \oplus N$. If ${}_SN_S = \oplus \sum Sb_i$ select $\{a_i\}_{i=1}^{t}$ from $\{p,b_j\}_{j=1}$ such that the images of the a_i determine a k-basis of m/m^2 . Let σ_i be the automorphism associated with the a_i for $1 \leq i \leq t$. We claim $R = S[a_1,\ldots,a_t;\sigma_1,\ldots,\sigma_t]$.

Let $T = S[a_1,\ldots,a_t;\sigma_1,\ldots,\sigma_t]$. Then $T \subseteq R$ and T maps surjectively on k . Let c be in R . Then there is a b in T with $c \equiv b \bmod m$.

If β is the nilpotency of m , construct $\{c_j\}_{j=0}^{\beta-1}$ satisfying

$$c \equiv c_j \bmod m^{j+1}$$

and

$$c_j \text{ is in } T .$$

This is done by first letting $c_0 = b$ and then if $j \geq 0$ note

$$c_j = c - \sum d_i w_i$$

where w_i is a power product of $\{a_i\}$. For each i , there is a d_i' in S with $d_i' \equiv d_i \bmod m$. Thus

$$c - c_j \equiv \sum d_i' w_i \bmod m^{j+2} .$$

Thus set $c_{j+1} = c_j + \sum d_i' w_i$. Since $m^\beta = 0$, $c = c_{\beta-1}$ and c is in $S[a_1,\ldots,a_t;\sigma_1,\ldots,\sigma_t]$ completing the proof.

(XIX.1) EXERCISE. (Clark). Let R be additively a vector space over $GF(2^2)$ with basis 1 and Θ . The multiplication in R is given by $\Theta\alpha = \alpha^2\Theta$ for α in $GF(2^2)$ and $\Theta^2 = 0$. Show the inner-automorphism

$$x \to uxu^{-1}$$

where $u = 1 + \Theta$ and $u^{-1} = 1 - \Theta$ takes the coefficient ring $GF(2^2)$ to another different, but isomorphic, coefficient ring.

(XIX.2) EXERCISE. (Wilson). Let R be a local ring with characteristic p^n and residue field $R/m = GF(p^t)$. Let $S = GR(p^n,t)$ and suppose that as an S-module

$$_S R = \oplus \sum_{i=1}^{q} Sb_i \ .$$

Show that R is the ring homomorphic image of a subring T of $M_q(S)$ where every entry below the main diagonal of each matrix in T is a multiple of p and every diagonal entry which is a pre-image of an element of m is also a multiple of p . (Hint: Use the technique of (XIX.2) with a careful choice of the independent S-generating set of $_S R$. In particular, if β is the nilpotency of m , consider all S-generating sets of $_S R$ as constructed in (XIX.4) with the maximal number of b_i's in $m^{\beta-1}$. Then, among those sets select the subsets with maximal number of additional b_i's in $m^{\beta-2}$. Continue.)

<u>(XIX.3)</u> <u>EXERCISE</u>. (It is not possible to improve (XIX.2) and ask that R be isomorphic to a subring of a matrix ring as this example of Wilson [137]illustrates.) Let $\Pi : Z/Z4 \to Z/Z2$ be the natural morphism and consider the set of matrices in both Z/Z4 and Z/Z2 of the form

$$\begin{bmatrix} a & b & \Pi(c) \\ 2d & a+2e & f \\ 2g & 2h & a+2i \end{bmatrix}$$

where a,b,c,... are in Z/Z4 . Define addition coordinate-wise and multiplication in the natural fashion always remembering to reduce the element in (1,3)-position modulo 2Z/Z4 as a final step.

(a) Show that these operations are well-defined and the set is a ring R .

(b) Show the ring R is local with characteristic 2^2 and

maximal ideal m consisting of all matrices whose elements on and below the diagonal are multiples of 2 .

(c) Show $R/m = GF(2)$.

(d) Consider the matrix

$$A = \begin{bmatrix} 0 & 0 & 1 \\ 0 & 0 & 0 \\ 2 & 0 & 0 \end{bmatrix}$$

in R . Show $2A = 0$. Thus, if R were a subring of $M_n(Z/4Z)$ then A would be in $2M_n(Z/4Z)$. But, if a matrix B is in $2M_n(Z/4Z)$ then $B^2 = 0$ while $A^2 \neq 0$. Deduce that R is not a subring of $M_n(Z/4Z)$.

(XIX.3) EXERCISE. Consider the decomposition of a local ring R in terms of its Galois ring S as given by (XIX.4):

$${}_S R_S = S \oplus N .$$

(a) Show that if the characteristic of R is p , then $N = m$ is an ideal of R and

$$R \cong (R/m) \oplus m .$$

(b) (Wilson). However, in general, N is not an ideal as this example illustrates. Let

$$R = \left\{ \begin{bmatrix} a & b \\ 2c & 2d \end{bmatrix} \mid a,b,c,d \text{ in } Z/4Z \right\} .$$

Show that R is local with maximal ideal

$$m = \left\{ \begin{bmatrix} 2a & b \\ 2c & 2d \end{bmatrix} \right\}.$$

Show that the Galois subring of R is

$$S = \left\{ \begin{bmatrix} a & 0 \\ 0 & a \end{bmatrix} \right\}$$

and thus $xSx^{-1} = S$ for all invertible x in R . Hence, Galois subrings of non-commutative rings may be unique absolutely. Every element of N is in m and thus has the above form. Since $\begin{bmatrix} 0 & 1 \\ 2 & 0 \end{bmatrix}$ is in R and $R = S \oplus N$, show there must be an element of N of the form $A = \begin{bmatrix} 2a & 1 \\ 2 & 2a \end{bmatrix}$. Show A^2 is in S and thus N is not an ideal.

(XIX.4) EXERCISE. (Raghavendran).

(*) (a) Let R be a local ring with characteristic p , maximal ideal m , and residue field $k = R/m$. Suppose $m^2 = 0$ and that $\dim_k(R) = n$. Show that R is isomorphic to the ring of all $n \times n$ matrices of the form

$$\begin{bmatrix} h_1 & f_2 & f_3 & \cdots & f_n \\ 0 & h_2 & 0 & \cdots & 0 \\ 0 & 0 & h_3 & \cdots & 0 \\ \cdot & \cdot & \cdot & & \\ 0 & 0 & 0 & \cdots & h_n \end{bmatrix}$$

where $h_1, f_2, \ldots, f_n$ are in k and $h_i = \sigma_i(h_1)$ for fixed automorphisms σ_i of k, $2 \le i \le n$. (Hint: Viewing R as a k-space select a skew-basis $b_2, \ldots, b_n$ of m. If Θ is a fixed cyclic generator of k^*, set $b_1 = \Theta$. Take $X = \text{diag}[\Theta, \sigma_2(\Theta), \ldots, \sigma_n(\Theta)]$ where $b_i\Theta = \sigma_i(\Theta)b_i$ and the elementary matrices $E_{12}, \ldots, E_{1n}$. For $r = \sum r_i b_i$ in R, define

$$r = \sum r_i b_i \to X^s + r_2E_{12} + \cdots + r_nE_{1n}$$

where $r = \Theta^s$ and $0 \to 0$.)

(b) Let R be a local ring with characteristic p, maximal ideal m, and residue field k. Suppose that $\dim_k(R) = n$ and $m^{n-1} \neq 0$.

(1) (Björk). Show that if b is an element of a skew-basis of m and b is not in m^2 then $b^{n-1} \neq 0$. (Hint: Assume $b^{n-1} = 0$. Show that there exist $f_1, \ldots, f_{n-1}$ in k and $z_1, \ldots, z_{n-1}$ in m^2 with

$$(f_1b + z_1) \cdots (f_{n-1}b + z_{n-1}) = 0 .$$

Deduce that $m^{n-1} = 0$.)

(**) (2) Show that R is isomorphic to the ring of n by n upper triangular matrices $[a_{ij}]$ over k satisfying

$$a_{i+1,j+1} = \sigma(a_{ij})$$

for σ an automorphism of k. (Hint: Use (i), the

matrix $Y = [y_{ij}]$ where $y_{ij} = 1$ if $j - i = 1$ and $y_{ij} = 0$ otherwise, and proceed in a manner similar to (a).)

(c) Generalize the above to the case $\chi(R) = p^{\lambda}$, $\lambda > 1$.

<u>(XIX.5)</u> <u>EXERCISE</u>. (Raghavendran). Let $k = GF(p^t)$ and $R = M_n(k)$. Show that R contains a subfield of order p^r if and only if r divides nt .

<u>(XIX.6)</u> <u>EXERCISE.</u> (Raghavendran). Let R be a ring containing a subfield $k = GF(q)$ where $q = p^t$. If $|R| = q^2$ show that R is isomorphic to one of the following types:

(a) $GF(q^2)$.

(b) If $t = 2w$, the ring of all 2×2 matrices over $GF(p^w)$.

(c) $GF(q) \oplus GF(q)$.

(d) The ring of 2×2 matrices of the form $\begin{bmatrix} a & b \\ 0 & \sigma(a) \end{bmatrix}$ with a and b in $GF(p^t)$ and σ is an automorphism of $GF(p^t)$.

<u>(XIX.7)</u> <u>EXERCISE</u>. (Raghavendran). Let R be a ring containing a subfield k of order $q = p^t$. If $|R| = q^3$ show that R is isomorphic to one of the following types:

(a) $GF(q^3)$.

(b) If $t = 3s$, then the ring of all 3 by 3 matrices over $GF(p^s)$.

(c) $GF(q^2) \oplus GF(q)$.

(d) $GF(q) \oplus GF(q) \oplus GF(q)$.

(e) If $t = 2s$, $M_2(GF(p^s)) \oplus GF(q)$.

(f) 2 by 2 upper triangular matrices over $GF(q)$.

(g) The ring (*) above with $n = 3$.

(h) The ring (**) above with $n = 3$.

(i) $GF(q) \oplus \left\{\begin{bmatrix} a & b \\ 0 & \sigma(a) \end{bmatrix} \mid a,b \text{ in } GF(q) \text{ and } \sigma \text{ a fixed automorphism of } GF(q)\right\}$.

<u>(XIX.8)</u> <u>EXERCISE</u>. (Ganesan and Raghavendran). Let R be a ring with $|R| = n^2 > 1$. If R has exactly n zero divisors show the following:

(a) $n = p^r$ for some prime p .

(b) The characteristic of R is p or p^2 .

(c) If the characteristic is p , then

$$R \simeq \left\{\begin{bmatrix} a & b \\ 0 & \sigma(a) \end{bmatrix} \mid a,b \text{ in } GF(p^r) \text{ and } \sigma \text{ a fixed automorphism of } GF(p^r)\right\} .$$

(d) If the characteristic of p^2 , then $R \simeq GR(p^2,2r)$.

<u>(XIX.9)</u> <u>EXERCISE</u>. (Raghavendran). Let p be a prime.

(a) Show that there are 4 non-isomorphic rings of p^2 elements. These are the <u>Singmaster</u> rings.

(b) Show that depending on whether p is even or odd there are respectively eleven or twelve mutually non-isomorphic rings of order p^3 .

<u>(XIX.10)</u> <u>EXERCISE</u>. (Quasi-Cyclic Rings). (Wirt). Let R be a local ring with characteristic p^n, maximal ideal m and Galois coefficient ring $S = GR(p^n,r)$. In the decomposition (XIX.4) $R = S \oplus N$, suppose that N is an ideal and N^2 is <u>complemented</u> in N; i.e., $N = N^2 \oplus T$ as (S,S)-modules.

(a) Show that if $n = 1$, this hypothesis is valid.

(b) Let $T^{(k)} = T \otimes_S \cdots \otimes_S T$ (k factors) for $1 \le k \le \beta - 1$ (β = nilpotency of m). Define

$$\Theta_k : T^{(k)} \to N$$

by

$$\Theta_k(t_1 \otimes \cdots \otimes t_k) = t_1 \cdots t_k .$$

Show Θ_k is an (S,S)-module morphism.

(c) Let $H = \oplus \sum_{i=1}^{\beta-1} T^{(k)}$. Extend the Θ_k to a unique (S,S)-module morphism

$$\sigma : H \to N$$

and show σ is surjective.

(d) Give H a ring structure (without identity) by defining a multiplication on generators as follows: For H_{k_1} in $T^{(k_1)}$ and H_{k_2} in $T^{(k_2)}$, define

$$H_{k_1} * H_{k_2} = \begin{cases} H_{k_1} \otimes H_{k_2} & \text{for } k_1 + k_2 \le m - 1 \\ 0 & \text{otherwise} \end{cases} .$$

Show H is a ring and $\sigma : H \to N$ is a ring morphism.

(e) Let $C = S \oplus H$ as (S,S)-modules. Give C a ring structure (without identity) by

$$(s_1,h_1)(s_2,h_2) = (s_1s_2, s_1h_2 + h_1s_2 + h_1 * h_2) .$$

Show that C is a ring.

(f) Define $\gamma : C \to R$ by $\gamma(s,h) = s + \sigma(h)$. Show that γ is a surjective ring morphism. Deduce that R is the image of

$$S \oplus H = S \oplus T^{(1)} \oplus \cdots \oplus T^{(\beta-1)} .$$

(g) Show that $T \simeq N/N^2$ and that $T^{(k)} = T \otimes \cdots \otimes T = T * \cdots * T = (T)^k$ is in C . Thus, show R is the homomorphic image of

$$S \oplus (N/N^2) \oplus \cdots \oplus (N/N^2)^{\beta-1} .$$

<u>(XIX.11)</u> <u>EXERCISE</u>. (An exercise on non-finite rings.) (Chew and Lawn). A ring R (not necessarily finite) is called <u>residually finite</u> if every proper homomorphic image is finite.

(a) Show Z , $GF(q)[X]$ and $GF(q)[[X]]$ are residually finite rings.

(b) Show that R is residually finite if and only if R satisfies the ascending chain condition and every non-zero prime ideal of R is of finite index.

(c) Show that R is residually finite if and only if $M_n(R)$ is residually finite.

(d) Let R be a commutative residually finite domain with quotient field k. Show

(1) Every subring S of k containing R is residually finite.

(2) If $a \neq 0$ is in R and $|R/Ra| = n$ then if S is a subring of k with $R \subseteq S \subseteq k$, $|S/Sa| \leq n^n$.

(XIX.12) EXERCISE. (Wilson, Clark).

(a) Let R be a local ring of characteristic p^λ. Show R is a Galois ring if and only if $Rad(R) = Rp$.

(b) Let R be a ring of characteristic p^λ. If $Rad(R) = Rp$, show R is a direct sum of matrix rings over Galois rings.

XX. THE SKEW-POLYNOMIAL RING $R[X;\sigma]$

In (XIX) a finite ring was characterized as a ring homomorphic image of a skew-polynomial ring. This short section concerns the simplest situation, that is, the skew-polynomial ring $R[X;\sigma]$ where R is a local ring and its elementary theory.

In this section R denotes a commutative local ring with maximal ideal m and residue field $k = R/m$. We examine the skew-polynomial ring $R[X;\sigma]$ where σ is an automorphism of R, X is an indeterminate and $rX = X\sigma(r)$ for all r in R. There is a natural ring morphism

$$\mu : R[X;\sigma] \to k[X;\bar{\sigma}]$$

where $\mu(X) = X$, $\mu(r) = r + m = \bar{r}$ and $\mu(\sigma) = \bar{\sigma}$ where $\bar{\sigma}$ is the automorphism of k given by $\bar{\sigma}(\bar{r}) = \overline{\sigma(r)}$. We appeal to the material in (II) for information on $k[X;\bar{\sigma}]$. Additional information on $k[X;\sigma]$ is available in Jacobson [15] and [67], Ore [93] and [95], and Cohn [8].

Until recently skew-polynomial rings have largely been utilized as a source of counter-examples. Ore [93] in 1933 considered skew-polynomial rings over fields and division rings developing results directly from the properties of the polynomials themselves. In [94] and [95] Ore specialized and improved his results when the coefficient ring was a finite field. Jacobson [15] in 1943 considered skew-polynomial rings over division rings as non-commutative Euclidean domains and in [67] examined cyclic algebras as homomorphic images of skew-polynomial rings. Cohn [8] has recently examined in greater generality the skew-polynomial ring over a division ring in attempting to provide a foundation for non-commutative algebraic geometry. Recently also Jategaonkar has explored structure theorems for skew-polynomial rings over semi-simple rings.

The first lemma follows from the morphism $\mu : R[X;\sigma] \to k[X;\bar{\sigma}]$.

<u>(XX.1)</u> <u>LEMMA</u>. Let $f = a_0 + a_1X + \cdots + a_nX^n$ be in $R[X;\sigma]$. Then

(a) f is nilpotent if and only if $a_0,\ldots,a_n$ are in m , i.e., $a_0,\ldots,a_n$ are nilpotent.

(b) f is a unit if and only if $a_1,\ldots,a_n$ are in m and a_0 is a unit.

Although (XX.1) is true for polynomial rings over arbitrary commutative rings, it fails for skew-polynomial rings over coefficient rings possessing non-trivial idempotents. An application of (XX.1) provides the following characterizations of the radical of $R[X;\sigma]$.

<u>(XX.2)</u> <u>THEOREM</u>. The following sets are equal.

(a) $\cup \{Q \mid Q$ is a two-sided nilpotent ideal of $R[X;\sigma]\}$.

(b) $\{f$ in $R[X;\sigma] \mid f^n = 0$ for some integer $n\}$.

(c) $\{f$ in $R[X;\sigma] \mid 1 + fg$ is a unit of $R[X;\sigma]$ for each g in $R[X;\sigma]\}$.

(d) $m[X;\sigma]$.

We denote any of the above sets by $\mathrm{Rad}(R[X;\sigma])$ or often use $m[X;\sigma]$ and call it the <u>radical</u> of $R[X;\sigma]$.

Let A be an ideal in $R[X;\sigma]$. Denote the image μA of A under μ by $\bar{A}$. By (II) $\bar{A} = (\bar{f})$ for some $\bar{f}$ in $k[X;\bar{\sigma}]$. Let f be a pre-image of $\bar{f}$ in $R[X;\sigma]$ and let $N = A \cap m[X;\sigma]$. Then the ideal $fR[X;\sigma] + N$ is contained in A . On the other hand, suppose g is in A . Then $\mu g = \bar{g} = \bar{f}\bar{h}$ in $k[X;\bar{\sigma}]$. If h is a pre-image of $\bar{h}$, then $g = fh + q$ for some q in $m[X;\sigma]$. Since $q = g - fh$, q is also in A and thus in $N = A \cap m[X;\sigma]$. Hence A is contained in $fS[X;\sigma] + N$.

<u>(XX.3)</u> <u>PROPOSITION</u>. Let A be an ideal in $R[X;\sigma]$. Then

$$A = fR[X;\sigma] + N$$
$$= R[X;\sigma]f + N$$

where $\mu A = (\mu f)$ and $N = A \cap m[X;\sigma]$.

The next lemma is straightforward. (See (II.11).)

(XX.4) LEMMA. (Division Algorithm.) Let f and g be polynomials in $R[X;\sigma]$. Assume the leading coefficient of g is a unit. Then there exist polynomials q and r in $R[X;\sigma]$ such that

$$f = gq + r$$

with $\deg(r) < \deg(g)$.

Similarly, one can find q' and r' with $f = q'g + r'$ and $\deg(r) < \deg(g)$. We define $\deg(0) = -\infty$. A polynomial f is called regular if $\mu f \neq 0$.

Adapting the proof of (XIII.5) (remember that $\sigma(m^i) \subseteq m^i$) we have the following theorem.

(XX.5) THEOREM. Let f be a regular polynomial in $R[X;\sigma]$. Then there exists a monic polynomial f^* with $\mu f = \mu f^*$ and $f = gf^*$ where g is a unit in $R[X;\sigma]$.

Thus the polynomial in (XX.3) may be taken to be monic. Indeed, f in (XX.3) may be chosen to be the monic polynomial of least degree in A .

(XX.1) EXERCISE. Let $f = a_0 + a_1X + \cdots + a_nX^n$ be in $R[X;\sigma]$.

(a) Show f is nilpotent if and only if f is a zero divisor if and only if $a_0, a_1, \ldots, a_n$ are in m .

(b) Part (a) is not in general true if the coefficient ring is not local. For a field k , let $S = k \oplus k$. Define $\sigma : S \to S$ by $\sigma\langle i,j\rangle = \langle j,i\rangle$. In $S[X;\sigma]$ show that $\langle 1,0\rangle X$ is nilpotent but $\langle 1,0\rangle$ is not nilpotent.

(c) Show f is a unit if and only if a_0 is a unit and $a_1, \ldots, a_n$ are in m .

(XX.2) EXERCISE. Use the above and prove (XX.2).

(XX.3) EXERCISE. Prove (XX.4).

(XX.4) EXERCISE. Show that every ideal in $R[X;\sigma]$ is finitely generated.

(XX.5) EXERCISE. Let $f, g, h, \ldots$ be non-zero polynomials in $R[X;\sigma]$.

(a) Show if f is a right (left) unit then f is a unit.

(b) Show that if $fg = f$ $(gf = f)$ then g is a unit.

(c) Show that 1 is the only idempotent in $R[X;\sigma]$.

(d) The polynomials f and g are *right associates* if $fR[X;\sigma] = gR[X;\sigma]$. Show that f and g are right associates if and only if $f = gh$ where h is a unit.

(e) If μf is irreducible, show f is irreducible.

(f) The polynomial f is *regular* if it is not a zero divisor.

Show f is regular if and only if $\mu f \neq 0$.

(XX.6) EXERCISE. Show that an ideal Q in $R[X;\sigma]$ is completely prime (maximal right), that is $R[X;\sigma]/Q$ is a finite field, if and only if

(a) $m[X;\sigma] \subseteq Q$

(b) μQ is completely prime (maximal right).

(XX.7) EXERCISE. (Factorization in $R[X;\sigma]$.) We assume a knowledge of the factorization in $k[X;\bar{\sigma}]$ as given in (II).

(a) Let f be a regular non-unit in $R[X;\sigma]$. Show

(1) $f = f_1 \dots f_n$ where the f_i are irreducible.

(2) If $\mu f = \bar{g}_1 \dots \bar{g}_m$ then $n \leq m$.

(b) (Hensel's Lemma for Galois Rings.) Let S be a Galois ring with maximal ideal S/Sp . Suppose f is in $S[X;\sigma]$ and

$$\mu f = \bar{g}_1 \dots \bar{g}_n$$

in $(S/Sp)[X;\sigma]$ where

(1) $\bar{g}_i(S/Sp)[X;\bar{\sigma}] = (S/Sp)[X;\bar{\sigma}]\bar{g}_i$

(2) $\bar{g}_i$ and $\bar{g}_j$, $i \neq j$, are coprime.

Show there exist $g_1,\dots,g_n$ in $S[X;\sigma]$ satisfying

(1) $\mu g_i = \bar{g}_i$, $1 \leq i \leq n$

(2) $f = g_1 \dots g_n$

(3) g_i and g_j , $i \neq j$, are coprime.

(Hint: Modify the proof of (XIII.4) and use the fact that $\sigma(p) = p$.)

(XX.8) EXERCISE. (Automorphisms of $S[X;\sigma]$ for S a Galois ring.)

(a) Let $\rho : S[X;\sigma] \to S[X;\sigma]$ be an S-automorphism. Show ρ is determined completely by its action on X. If $\sigma \neq i_S$, show that $\rho(X) = uX$ for some unit u of S. Conversely, if $\rho : S[X;\sigma] \to S[X;\sigma]$ is an S-morphism and $\rho(X) = uX$ where u is a unit, then ρ is an automorphism.

(b) Let $\phi : S \to S$ be a ring morphism and u be in S. Define

$$\phi_u : S[X;\sigma] \to S[X;\sigma]$$

by

$$\phi_u(g) = \phi(a_0) + \phi(a_1)(uX)^1 + \cdots + \phi(a_n)(uX)^n$$

for $g = a_0 + a_1X + \cdots + a_nX^n$. Show ϕ_u is a ring automorphism if ϕ is an automorphism and u is a unit.

(XX.9) EXERCISE. (Chain Rings.) (See (XVII).) A ring R is called a left (right) chain ring if its lattice of left (right) ideals forms a chain.

(a) (Clark.) Show the following are equivalent:

(1) R is a left chain ring.

(2) R is a right chain ring.

(3) R is a local ring with maximal ideal $m = R\Theta$ for any Θ in $m - m^2$.

(4) The principal left ideals of R form a chain.

(b) (Clark and Drake.) Let R be a chain ring with characteristic p^n, maximal ideal $m = R\Theta$ and Galois coefficient ring

$S = GR(p^n, r)$. Let β denote the nilpotency of m . Show there exist integers r and t satisfying

(1) $R = S \oplus S\Theta \oplus \cdots \oplus S\Theta^{r-1}$ as an (S,S)-module where $s\Theta^i = \Theta^i\sigma^i(s)$ for some fixed automorphism σ of S and s in S .

(2) $\Theta^r = p(s_{r-1}\Theta^{r-1} + \cdots + s_1\Theta + s_0)$ where s_i are in S and s_0 is a unit.

(3) There are (S,S)-module isomorphisms

$$S\Theta^i \simeq S \qquad 1 \le i \le t - 1$$
$$S\Theta^i \simeq Sp \qquad t \le i \le r - 1 .$$

(4) $\beta = (n - 1)r + t$, $1 \le t \le r$, where r is the greatest integer $\le \beta$ with p in m^r .

(Hint: See (XVII.4).)

(c) (Wirt, Clark, Drake.) Assume the hypothesis of (b). Show that

$$R \simeq S[X;\sigma]/(g, p^{n-1}X^t)$$

where $t = \beta - (n - 1)r > 0$ and g is an Eisenstein polynomial (see (XVII)) of degree r over S .

(XX.10) EXERCISE. (One-Step Rings.) (Wirt, Redei.) A ring is a one-step ring if every proper subring is commutative.

(a) If R is a one-step ring, show that R is local.

(b) Let p and q be primes and m,e,n positive integers with $n < q$. Let S denote the Galois ring $GR(p^m, q^e)$.

By (XV.11) select a separable generator a of S over Z/Zp^m such that $Aut(S)$ are given as power maps on this generator. Let $t = p^{nq^{e-1}}$ and define $\sigma : S \to S$ by $\sigma : a \to a^t$. Show

$$S[X;\sigma]/(X^2)$$

is a one-step ring. Further, every one-step ring has this form. (Hint: see Redei, Algebra, p.753, Theorem 447, Ring R_{II} .)

(XX.11) EXERCISE. Let S be a Galois ring of characteristic p^n and A an ideal in $S[X;\sigma]$ with $A \cap S = 0$.

(a) Consider the chain

$$A = A_0 \subseteq A_1 \subseteq \cdots \subseteq A_{n-1} \subseteq A_n = S[X;\sigma]$$

where $A_i = \{f \text{ in } S[X;\sigma] \mid p^i f \text{ is in } A\}$. Show these ideals are mapped onto the ascending chain of submodules

$$0 \subseteq B_1 \subseteq B_2 \subseteq \cdots \subseteq B_{n-1} \subseteq B_n = B$$

of the S-module $B = S[X;\sigma]/A$ where

$$B_i = \{r \text{ in } B \mid p^i r = 0\} .$$

(b) Let f_i denote the monic polynomial of least degree in A_i and $N_i = A_i \cap (Sp)[X;\sigma]$. Thus $f = f_0$, $N = N_0$, $f_n = 1$, $N_n = (Sp)[X;\sigma]$ and for $0 \le i \le n$, $A_i = (f_i,N_i)$. Here (f_i,N_i) denotes the ideal $f_iS[X;\sigma] + N_i$. Let $h_i = \deg(f_i) - \deg(f_{i-1})$

for $i = 1,2,\ldots,n$. Show the S-module $B = S[X;\sigma]/A$ is a direct sum of

$$h = h_1 + h_2 + \cdots + h_n$$

cyclic S-modules which are generated by the residue classes of $f_i, f_iX, \ldots, f_iX^{h_i-1}$ for $1 \le i \le n$ modulo A .

(c) Show $A = (f, pf_1, p^2f_2, \ldots, p^{n-1}f_{n-1})$.

(d) Show the S-module $S[X;\sigma]/A$ is S-free if and only if $A = (f)$.

(XX.12) EXERCISE. Show every ideal in $R[X;\sigma]$ is finitely generated. Show every ideal in $R[X_1,\ldots,X_n;\sigma_1,\ldots,\sigma_n]$ is finitely generated. (Hint: Use a proof similar to proof of Hilbert Basis Theorem.)

XXI. THE UNITS OF A FINITE RING

> This concluding section examines the question of the structure of the group of units of a finite ring. In general, there is little known. In this section we examine the basic results.

We again ask about the structure of the group of units in a finite ring. As indicated in (II) the units of a finite field form a cyclic group. The group of units of Z/Zp^n is well-known, occurs in the exercises in (I) and may be deduced from the structure of the units of the Galois rings given in (XVII). The units of the matrix ring over a finite field were examined in (VIII). Finally, (XVIII) provided some information on the units of local commutative rings. This section and its exercises will provide certain basic facts concerning the structure of the units in an arbitrary finite ring. In general, little is known.

Let R^* denote the group of units of a ring R. If R decomposes as a direct sum $R = R_1 \oplus \cdots \oplus R_n$ of rings R_i then R^* decomposes naturally as a direct product $R^* = R_1^* \times \cdots \times R_n^*$ of groups. Thus we generally will assume R is indecomposable.

We begin by examining ring morphisms on R and their effect on R^*.

If R and S are rings and $\sigma : R \to S$ is a ring morphism then σ restricted to R^*, $\sigma|_{R^*}$, is a group morphism of R^* into S^*. We will denote the restriction of σ to R^* also by σ.

The proof of the following lemma is easy.

(XXI.1) LEMMA. An element of a ring R is a unit if and only if it is a unit modulo $\mathrm{Rad}(R)$.

(XXI.2) THEOREM. Let K be a left ideal of a ring R. Suppose an element r in R satisfies $K + Rr = R$. Then $K + r$ contains a unit of R.

Proof. (Bass) By the lemma an element of R is invertible if and only if it is invertible modulo $\mathrm{Rad}(R)$. This reduces the case to $R/\mathrm{Rad}(R)$. Hence assume $\mathrm{Rad}(R) = 0$. Since $R = \oplus \sum R_i$ of simple rings R_i, it suffices to consider each summand. Thus assume $R = \mathrm{End}_k(V)$ where V is a finite dimensional vector space over a finite field k.

Then K is the set of all $\alpha : V \to V$ with $\alpha V \subset KV = W$. Since $K + Rr = R$, $W + \mathrm{Im}(r) = V$. Select $W_0 \subset W$ with $V = W_0 \oplus \mathrm{Im}(r)$. If U is chosen so that $V = \ker(r) \oplus U$ then r induces an isomorphism $U \to \mathrm{Im}(r)$. Select α with $\alpha U = 0$ and such that α induces an isomorphism $\ker(r) \to W_0$. Then $\alpha V = W_0 \subset W$ so α is in K. Further $\alpha + r$ is an automorphism of V.

(XXI.3) COROLLARY. Let $\sigma : R \to S$ be a surjective ring morphism. Then $\sigma : R^* \to S^*$ is a surjective group morphism.

Proof. Let $K = \ker(\sigma)$ where $\sigma : R \to S$. If s in S is a unit and $\sigma(r) = s$, then $K + Rr = R$. By the theorem $K + r$ contains a unit $u = t + r$ with t in K . Then $\sigma(u) = s$.

Observe if $p \neq 2$ is a prime, then the natural surjective morphism $Z \to Z/Zp$ does not carry $Z^* = \{1,-1\}$ onto $(Z/Zp)^*$.

The above corollary permits the use of the Galois ring and regular representations of R in the study of R^* .

(XXI.4) THEOREM. Let R be a ring with Galois coefficient ring S . If, as a left S-module,

$$ {}_S R = \oplus \sum_{i=1}^{t} Sb_i \, , $$

then R^* is the homomorphic image of a subgroup of the general linear group $GL_t(S)$ over S .

Proof. Apply (XIX.2) to (XXI.3).

We now examine $R \to R/\mathrm{Rad}(R)$. The kernel of this ring morphism is $\mathrm{Rad}(R)$ — a nilpotent ideal.

(XXI.5) THEOREM. (Ayoub [142]) Let N be a nilpotent ideal of R. Then

(a) $H = 1 + N$ is a normal subgroup of R^*.

(b) R^*/H is isomorphic to $(R/N)^*$.

(c) For each integer $i \geq 1$, the additive group N^i/N^{i+1} is isomorphic to the multiplicative group $(1 + N^i)/(1 + N^{i+1})$.

Proof. Properties (a) and (b) follow easily. We prove (c).

Since N^i and N^{i+1} are ideals, $1 + N^i$ and $1 + N^{i+1}$ are normal subgroups of G and $1 + N^{i+1}$ is normal in $1 + N^i$. Define

$$\eta : N^i \to (1 + N^i)/(1 + N^{i+1})$$

by

$$\eta(x) = (1 + x)(1 + N^{i+1}) .$$

Let x and y be in N^i. Since $1 + N^i$ is a multiplicative group select z in N^i with $(1 + x + y)(1 + z) = 1$. Then

$$(1 + x)(1 + y) = (1 + x + y)(1 + (1 + z)xy) .$$

Thus

$$[(1 + x)(1 + N^{i+1})][(1 + y)(1 + N^{i+1})] = (1 + x + y)(1 + N^{i+1}) ,$$

that is, $\eta(x)\eta(y) = \eta(x + y)$ and η is a group morphism. The kernel of η is $\{x \text{ in } N^i \mid 1 + x \text{ is in } 1 + N^{i+1}\} = N^{i+1}$ and we are done.

A special case involving $1 + N$ is worthy of note.

Suppose a ring R contains a subring S which is a direct sum of matrix rings over local rings and that R as an S-module decomposes

$$R = S \oplus N$$

where N is in the radical of R. For example, S may be a direct sum of matrix rings over Galois rings as in (XIX.5) or we may have the standard decomposition for R (VIII.27).

If $N^2 \subset N$ then N is an ideal of R.

(XXI.6) THEOREM. (For the above setting.) If $N^2 \subset N$ then R^* is the semi-direct product of S^* and $1 + N$.

Proof. Since $N^2 \subset N$, $1 + N$ is a normal subgroup of R^*. Clearly $S^*(1 + N) = R^*$ and $S^* \cap (1 + N) = 1$. Thus R^* is a semi-direct product of S^* and $1 + N$ (for example, see Rotman [19]). This can be made more explicit as follows. Since $R^* = S^*(1 + N)$ each element of R^* may be written uniquely as a product sh where s is in S^* and h is in $1 + N$. Since $1 + N$ is normal, $s^{-1}hs$ (which we denote by h^s) is in $1 + N$. For the product of two elements $s_1h_1s_2h_2 = s_1s_2(s_2^{-1}h_1s_2)h_2 = s_1s_2h_1^{s_2}h_2$. That is, identifying R^* with the set $\{\langle s,h\rangle \mid s \text{ in } S \text{ and } h \text{ in } 1 + N\}$, multiplication is given by

$$\langle s_1,h_1\rangle\langle s_2,h_2\rangle = \left\langle s_1s_2, h_1^{s_2}h_2\right\rangle$$

where $h_1^{s_2} = s_2^{-1}h_1s_2$.

(XXI.7) COROLLARY. Let R be a ring of characteristic p. Then R^* is a semi-direct product of $(R/Rad(R))^*$ and $1 + Rad(R)$.

Proof. If the characteristic of R is p, then R is an algebra over a finite field. Applying the Wedderburn Principal Theorem (VIII.28), $R = S \oplus Rad(R)$ where $S \simeq R/Rad(R)$. Now apply the above.

Let G be a group written multiplicatively. For x and y in G, $x^{-1}y^{-1}xy$ is called the commutator of x and y and is denoted by $[x,y]$. The commutator subgroup of G, denoted $[G,G]$ or G', is the subgroup of G generated by all the commutators in G. Higher commutator subgroups of G are defined inductively by

$$G^{(0)} = G \quad \text{and} \quad G^{(i+1)} = [G^{(i)},G^{(i)}]$$

for $i \geq 0$.

Recall a group G is solvable if $G^{(n)} = 1$ for some integer n. Equivalently, G is solvable if G has a composition chain with cyclic factor groups of prime order. Define a second chain of subgroups of G, denoted $\delta_i(G)$, as follows:

$$\delta_1(G) = G, \quad \delta_{i+1}(G) = [\delta_i(G),G].$$

Recall a group G is nilpotent if $\delta_n(G) = 1$ for some integer n. A direct computation provides the following result.

(XXI.8) COROLLARY. Let N be a nilpotent ideal of R. Then $1 + N$ is a nilpotent group.

The nilpotent groups form a class of groups lying strictly between the Abelian groups and the solvable groups. The kernel of $R^* \to (R/Rad(R))^*$ is $1 + Rad(R)$ and is thus nilpotent. However, R^* is rarely nilpotent or even solvable, as the next result indicates.

(XXI.9) THEOREM. (Eldridge) Let R be a ring. Then R^* is solvable if and only if the direct summands of $R/Rad(R)$ are finite fields or isomorphic to the ring of 2 by 2 matrices over $Z/Z2$.

Proof. Recall if G is a group with normal subgroup N then G is solvable (nilpotent) if and only if G/N and N are solvable (nilpotent). We have $R/Rad(R) = \oplus \sum M_{n_i}(k_i)$ and $R^*/(1 + Rad(R)) = GL_{n_1}(k_1) \times \cdots \times GL_{n_t}(k_t)$. But $GL_n(k)$ is solvable only if $n = 1$ or $n = 2$ and $k = Z/Z2$.

(XXI.10) COROLLARY. Let R be a ring.

(a) R^* is nilpotent if and only if $R/Rad(R)$ is a direct sum of fields.

(b) If R is a basic ring then R^* is nilpotent.

We now provide one result concerning the elements of R^*.

Let α and β be in R^* where R is a ring. Write $\alpha \sim \beta$ if there is a unit u in R^* with $u^{-1}\alpha u = \beta$, i.e., $\alpha \sim \beta$ if α and β differ by an inner-automorphism. Let $|\alpha|$ denote the order of α in R^* .

(XXI.11) THEOREM. (Pomfret) Let R be a ring of characteristic $\chi(R) = p^{\lambda}$ for p a prime. Let α and β be in R^* with $(|\alpha|,p) = 1$ and $(|\beta|,p) = 1$. Let $\bar{\alpha}$ and $\bar{\beta}$ be the images of α and β in $R/\text{Rad}(R)$. Then $\alpha \sim \beta$ if and only if $\bar{\alpha} \sim \bar{\beta}$.

Proof. Since units go to units modulo Rad(R) , $\alpha \sim \beta$ implies $\bar{\alpha} \sim \bar{\beta}$.

To prove the converse, let $N = \text{Rad}(R)$. Then $H = 1 + N$ is a solvable p-group of order, say, $|H| = p^m$. Let $\langle\alpha\rangle$ denote the group generated by α . Since H is normal, $\langle\alpha\rangle H$ is a subgroup of R^* . Further, H normal in $\langle\alpha\rangle H$, H solvable and $\langle\alpha\rangle$ solvable imply $\langle\alpha\rangle H$ is solvable.

Since $\bar{u}^{-1}\bar{\beta}\bar{u} = \bar{\alpha}$ for some u in R^* it follows that $\langle\alpha\rangle H = \langle u^{-1}\beta u\, H\rangle$. Since $\langle\alpha\rangle H$ is solvable, we apply Hall's Theorem ([12], Thm. 9.3.1, p.141) and conclude that $\langle\alpha\rangle$ and $\langle u^{-1}\beta u\rangle$ are conjugate in $\langle\alpha\rangle H$, i.e., differ by an inner-automorphism. Thus there is a δ in H with $\delta^{-1}u^{-1}\beta u\delta = \alpha^t$ for some t . Then $\bar{\alpha} = \bar{u}^{-1}\bar{\beta}\bar{u} = \bar{\alpha}^t$. Since α and $\bar{\alpha}$ have the same order, $\alpha = \alpha^t$. Thus $\alpha = (u\delta)^{-1}\beta(u\delta)$ and $\alpha \sim \beta$.

(XXI.1) EXERCISE. It is natural to ask, "Do the units R^* of a ring R generate R ?" If $S \subset R$ let $[S]$ denote the subring generated by S in R.

(a) If $R = (Z/Z2) \oplus (Z/Z2)$, show $[R^*] \neq R$.

(b) (Stewart [163]) Part (a) provides the pathological case. Let $R = R_1 \oplus \cdots \oplus R_n$ where the R_i are indecomposable rings. Show that if no two summands of R equal $Z/Z2$ then $[R^*] = R$.

(Hint: First consider $R = M_n(k)$.)

(XXI.2) EXERCISE. Let R be a finite ring with $N = \mathrm{Rad}(R)$.

(a) Show that N is a group under $\circ$ where $x \circ y = x + y + xy$. Denote this group by R^0.

(b) If $R = R_1 \oplus \cdots \oplus R_n$ show $R^0 = R_1^0 \times \cdots \times R_n^0$.

(c) Show $R^0 \simeq R^*$.

(XXI.3) EXERCISE. (Eldridge) If R^* is a simple group, show R has one of the following forms:

(a) $(Z/Z2) \oplus \cdots \oplus (Z/Z2)$.

(b) $GF(3)$.

(c) $GF(2^m)$ where $2^m - 1$ is a prime.

(XXI.4) EXERCISE. Show the only non-commutative ring with a cyclic group of units is

$$\left\{ \begin{bmatrix} x & y \\ 0 & z \end{bmatrix} \mid x,y,z \text{ in } Z/Z2 \right\}.$$

(XXI.5) EXERCISE. (Farahat) Let e be an idempotent in a ring R. Show there is a unique maximal multiplicative subgroup $G_e(R)$ of R with identity e. Show $G_e(R) = (eRe)^*$.

(XXI.6) EXERCISE. (Farahat) Let $\delta(R) = |R^*|/|R|$.

(a) If $\sigma : R \to S$ is a surjective ring morphism, show $\delta(R) = \delta(S)$.

(b) Let k be a field of q elements. Show

$$\delta(M_n(k)) = (1 - q^{-1})(1 - q^{-2})\cdots(1 - q^{-n}) .$$

(c) Let R be a ring. Show

$$\delta(R) = \prod_{n,q} (1 - q^{-1})(1 - q^{-2})\cdots(1 - q^{-n})$$

where the product extends over all pairs $\langle n,q \rangle$ corresponding to summands $M_n(k)$, $|k| = q$, in $R/\mathrm{Rad}(R)$.

(d) If $R = R_1 \oplus \cdots \oplus R_n$ show $\delta(R) = \prod_i \delta(R_i)$.

(XXI.7) EXERCISE. (Eldridge and Fischer [149]) Let R be a ring. Let R^* be Abelian. Show that if $R/\mathrm{Rad}(R)$ is a direct sum of finite fields, none of order 2, then R is commutative. Use this and show, if R^* is Abelian and 2 is a unit in R then R is commutative.

(XXI.8) EXERCISE. It is difficult to examine the normal subgroups of R^* when R decomposes as a direct sum.

(a) Give an example of a group $G \times H$ containing a normal non-trivial subgroup N such that $N \cap G = 1$ and $N \cap H = 1$.

Conclude that $N \neq (N \cap G) \times (N \cap H)$.

(b) Let N be a subgroup of $G \times H$. Let N_1 be the projection of N on G and $F_1 = \{g \mid g$ is in G and $\langle g,1 \rangle$ is in $N\}$. Let N_2 and F_2 be defined similarly. Show F_i is normal in N_i and $N_1/F_1 \simeq N_2/F_2$. Show that if N is normal in $G \times H$ then F_i is normal in $G \times H$ and N_1/F_1 is in the center of G/F_1 (similarly N_2/F_2 is in the center of H/F_2).

(XXI.9) EXERCISE. (Ditor [146]) This is a surprising result by Ditor.

(a) Show a finite group G of odd order is the group of units of some ring if and only if G is Abelian and is the finite direct product of cyclic groups G_i where $|G_i| = 2^{K_i} - 1$.

(b) If G is the group of units of a ring R and G is finite of odd order, show the subring $[G]$ generated by G in R is a direct sum of finite fields of characteristic 2 .

[See S. Z. Ditor [146].]

(XXI.10) EXERCISE. (On non-finite rings.) Show an Artinian ring with a finite group of units is a finite ring.

BIBLIOGRAPHY

A. Books

[1] A. Adrian Albert, Fundamental Concepts of Higher Algebra, Univ. of Chicago Press, Chicago, 1956.

[2] ______________, Structure of Algebras, Amer. Math. Society, Providence, R. I., 1961.

[3] E. Artin, C. Nesbitt and R. Thrall, Rings with Minimum Condition, Univ. of Michigan Press, Ann Arbor, 1944.

[4] H. Bass, Algebraic K-Theory, Benjamin, New York, 1968.

[5] E. R. Berlekamp, Algebraic Coding Theory, McGraw-Hill, New York, 1968.

[6] R. Brauer, Non-Commutative Rings, Harvard Univ. Press, Cambridge, Mass., 1966.

[7] S. Chase, D. Harrison and A. Rosenberg, Galois Theory and Cohomology of Commutative Rings, Amer. Math. Soc., Memoirs #52, Providence, R. I., 1965.

[8] P. M. Cohn, Free Rings and Their Relations, Academic Press, New York, 1971.

[9] C. Cullen, Matrices and Linear Transformations, Addison-Wesley, Reading, Mass., 1967.

[10] L. Dickson, Linear Groups with an Exposition of the Galois Field Theory, Dover (Reprint), New York, 1958.

[11] J. Dixon, The Structure of Linear Groups, Van Nostrand, Math. Studies #37, New York, 1971.

[12] M. Hall, Jr., The Theory of Groups, MacMillan, New York, 1959.

[13] K. Hofmann (Ed.), Lectures on Rings and Modules, Springer-Verlag, Lecture Notes in Math. #246, New York, 1972.

[14] F. DeMeyer and E. Ingraham, Separable Algebras over Commutative Rings, Springer-Verlag, Lecture Notes in Math. #181, New York, 1971.

[15] N. Jacobson, The Theory of Rings, Amer. Math. Soc., Math. Surveys #II, Providence, R. I., 1943.

[16] __________, The Structure of Rings, Amer. Math. Soc., Colloquium Pub. Vol. 37, Providence, R. I., 1964.

[17] R. Kruse and D. Price, Nilpotent Rings, Gordon and Breach, New York, 1970.

[18] M. Nagata, Local Rings, Wiley-Interscience, New York, 1962.

[19] J. Rotman, The Theory of Groups, Allyn and Bacon, Boston, Mass., 1965.

B. Artinian Rings

[20] M. Auslander, "Notes on Representation Theory of Artin algebras," (preprint: written by Idun Reiten).

[21] M. Auslander and I. Reiten, "Stable equivalence of Artin algebras," preprint.

[22] H. Bass, "Finitistic homological dimension and a homological generalization of semi-primary rings," Trans. A.M.S. 95(1960), 466-488.

[23] S. Brenner, "Large indecomposable modules over a ring of 2×2 triangular matrices," Bull. London Math. Soc. 3(1971), 333-336.

[24] R. Courter, "Finite direct sums of complete matrix rings over perfect completely primary rings," Canad. J. Math. 21(1969), 430-446.

[25] C. Curtis, "The structure of non-semisimple algebras," Duke Math. J. 21(1954), 79-85.

[26] S. Endo, "Completely faithful modules and quasi-Frobenius algebras," J. Math. Soc. Japan 19(1967), 437-456.

[27] K. Fuller, "On indecomposable injectives over Artinian rings," Pacific J. Math. 29(1969), 115-135.

[28] ________, "Double centralizers of injectives and projectives over Artinian rings," Illinois J. Math. 14(1970), 658-664.

[29] J. Golan, "Characterization of rings using quasiprojective modules," Israel J. Math. 8(1970), 34-38.

[30] M. Hall, "The position of the radical in an algebra," Trans. A.M.S. 48(1940), 391-404.

[31] T. Hannula, "On the construction of quasi-Frobenius rings," J. of Algebra 25(1973), 403-414.

[32] M. Harada, "Hereditary semi-primary rings and tri-angular matrix rings," Nagoya Math. J. 26-27(1966), 463-484.

[33] A. Heller and I. Reiner, "Indecomposable representations," Illinois J. Math. 5(1961), 314-323.

[34] G. Hochschild, "Automorphisms of simple algebras," Trans. A.M.S. 69(1950), 292-301.

[35] M. Menzin, Indecomposable Modules over Artin Local Rings, Thesis: Brandeis University, 1970.

[36] K. Morita, "Duality for modules and its application to the theory of rings with minimum condition," Tokyo Kyoiku Daigaku, Science Reports 6(1958), 83-142.

[37] ________, "Category-isomorphisms and endomorphism rings of modules," Trans. A.M.S. 103(1962), 451-469.

[38] K. Morita, Y. Kawada and H. Tachikawa, "On injective modules," Math. Zeitschr. 68(1957), 217-226.

[39] K. Morita and H. Tachikawa, "Character modules, submodules of a free module and quasi-Frobenius rings," Math. Zeitschr. 65(1956), 414-428.

[40] C. Nesbitt, "On the regular representation of algebras," Annals of Math. 39(1938), 634-658.

[41] M. Osima, "Notes on basic rings," Math. J. Okayama Univ. 2(1953), 103-110.

[42] ________, "Notes on basic rings II," Math. J. Okayama Univ. 3(1954), 121-127.

[43] H. Tachikawa, "Duality theorem of character modules for rings with minimum condition," Math. Zeitschr. 68(1958), 479-487.

[44] B. Vinograde and G. P. Weeg, "Maximally uncleft rings and algebras," Illinois J. Math. 3(1959), 272-284.

[45] J. Walter, "Structure of cleft rings, I," Illinois J. Math. 3(1959), 445-467.

[46] ________, "Structure of cleft rings, II," Illinois J. Math. 4(1960), 376-396.

[47] R. Warfield, "A Krull-Schmidt theorem for infinite sums of modules," Proc. A.M.S. 22(1969), 460-465.

[48] __________, "Purity and algebraic compactness for modules," Pacific J. Math. 28(1969), 699-719.

C. Local Rings

[49] G. Azumaya, "On maximally central algebras," Nagoya Math. J. 2(1951), 119-150.

[50] E. Batho, "Non-commutative semi-local and local rings," Duke Math. J. 24(1957), 163-172.

[51] I. S. Cohen, "On the structure and ideal theory of complete local rings," Trans. A.M.S. 59(1946), 54-106.

[52] I. S. Cohen and I. Kaplansky, "Rings for which every module is a direct sum of cyclic modules," Math. Zeitschr. 54(1951), 97-101.

[53] T. Hungerford, "On the structure of principal ideal rings," Pacific J. Math. 25(1968), 543-547.

[54] W. Krull, "Algebraische theorie der ringe," Math. Annalen 92 (1924), 183-213.

[55] H. Lenzing, "A homological characterization of Steinitz rings," Proc. A.M.S. 29(1971), 269-271.

[56] K. McLean, "Commutative Artinian principal ideal rings," Proc. London Math. Soc. 26(1973), 249-272.

[57] E. Snapper, "Completely primary rings, I," Annals of Math. 52(1950), 666-693.

[58] ________, "Completely primary rings, II," Annals of Math. 53(1951), 125-142.

[59] ________, "Completely primary rings, III," Annals of Math. 53(1951), 207-234.

[60] ________, "Completely primary rings, IV," Annals of Math. 55(1952), 46-64.

D. Polynomial Rings

[61] J. Arnold and R. Gilmer, "On the contents of polynomials," Proc. A.M.S. 24(1970), 556-562.

[62] D. Coleman and E. Enochs, "Isomorphic polynomial rings," preprint.

[63] B. Elkins, "Characterization of separable ideals," Pacific J. Math. 34(1970), 45-49.

[64] E. Feller and E. Swokowski, "On ring extensions for completely primary noncommutative rings," Trans. A.M.S. 105(1962), 251-263.

[65] D. Fields, "Zero divisors and nilpotent elements in power series rings," Proc. A.M.S. 27(1971), 427-433.

[66] R. Gilmer, "R-automorphisms of R[X] ," Proc. London Math. Soc. 18(1968), 328-336.

[67] N. Jacobson, "Non-commutative polynomials and cyclic algebras," _Annals of Math._ 35(1934), 197-210.

[68] G. Janusz, "Separable algebras over commutative rings," _Trans. A.M.S._ 122(1966), 461-479.

[69] R. MacRae, "On the notion of separable ideals," _Archiv. der Math._ 18(1967), 591-594.

[70] T. Nagahara, "Characterization of separable polynomials over a commutative ring," _Proc. Japan Academy_ 46(1970), 1011-1015.

[71] O. Ore, "Theory of non-commutative polynomials," _Annals of Math._ 34(1933), 480-508.

E. Finite Fields and Their Matrix Rings and Polynomial Rings

[72] J. Ax, "Zeros of polynomials over finite fields," _Amer. J. Math._ 86(1965), 225-261.

[73] E. Berlekamp, "Distribution of cyclic matrices in a finite field," _Duke Math. J._ 33(1966), 45-48.

[74] ___________, "Factoring polynomials over large finite fields," _Computation_ 24(1970), 713-735.

[75] J. Brawley, "Enumeration of canonical sets by rank," _Amer. Math. Monthly_ 74(1967), 176-177.

[76] _________, "Polynomial representation of functions in a finite field," Technical Report #94, Dept. of Math., Clemson Univ., 1973.

[77] J. Brawley, L. Carlitz and J. Levine, "Power sums of matrices over a finite field," Technical Report #143, Dept. of Math., Clemson Univ., 1973.

[78] J. Brawley, L. Carlitz and T. Vaughn, "Linear permutation polynomials with coefficients in a subfield," preprint.

[79] L. Carlitz, "The arithmetic of polynomials in a Galois field," Amer. J. Math. 54(1932), 39-50.

[80] __________, "Invariant theory of systems of equations in a finite field," J. D'Analyse Math. 3(1953/54), 382-413.

[81] L. Carlitz and J. Hodges, "Distribution of matrices in a finite field," Pacific J. Math. 6(1956), 225-230.

[82] D. Carlson, "A note on matrices over extension fields," Duke Math. J. 33(1966), 503-505.

[83] W. Feit and N. Fine, "Pairs of commuting matrices over a finite field," Duke Math. J. 27(1960), 91-94.

[84] N. Fine and I. Herstein, "The probability that a matrix be nilpotent," Illinois J. Math. 2(1958), 499-504.

[85] S. Fisher and M. Alexander, "Matrices over a finite field," Amer. Math. Monthly 73(1966), 639-641.

[86] R. Fray and R. Gilmer, "On solvability by radicals of finite field," Math. Annalen 199(1972), 279-291.

[87] M. Fredman, "Congruence formulas obtained by counting irreducibles, Pacific J. Math. 35(1970), 613-624.

[88] J. Goldman and Gian-Carlo Rota, "The number of subspaces of a vector space," Recent Progress in Combinatorics 75-83, Academic Press, New York, 1969.

[89] ______________________________, "On the foundations of combinatorial theory IV, Finite vector spaces and Eulerian generating functions," Studies in Applied Math. XLIX(1970), 239-258.

[90] J. Hodges, "Scalar polynomial equations for matrices over a finite field," *Duke Math. J.* 25(1958), 291-296.

[91] A. Long, "Factorization of irreducible polynomials over a finite field with the substitution $X^{p^r} - X$ for X," *Duke Math. J.* 40(1973), 63-76.

[92] J. MacWilliams, "Orthogonal matrices over finite fields," *Amer. Math. Monthly* 76(1969), 152-164.

[93] O. Ore, "Theory of non-commutative polynomials," *Annals of Math.* 34(1933), 480-508.

[94] ______, "On a special class of polynomials," *Trans. A.M.S.* 35 (1933), 569-584.

[95] ______, "Contributions to the theory of finite fields," *Trans. A.M.S.* 36(1933), 243-275.

[96] W. Peterson, "Some new results on finite fields and their application to the theory of BCH Codes," 329-334, *Combinatorial Theory and Its Applications*, Ed. R. Bose and T. Dowling, North Carolina Press, 1971.

[97] R. Schafer, "A remark on finite simple rings," *Amer. Math. Monthly* 60(1953), 696-697.

[98] B. Stewart, "The class equation with respect to row-equivalence of matrices with elements in a Galois field," preprint.

[99] J. Thas, "The m-dimensional projective space $S_m(M_n(GF(q)))$ over the total matrix algebra $M_n(GF(q))$ of $n \times n$-matrices with elements in the Galois field $GF(q)$," *Rendiconti di Math.* 4(1971), 459-532.

[100] G. Walker, "Fermat's theorem for algebras," Pacific J. Math. 4(1954), 317-320.

[101] T. Vaughn, Polynomials and Linear Transformations over Finite Fields, Thesis: Duke University, 1972.

F. Finite Rings

[102] D. Bollman and H. Ramirez, "On the number of nilpotent matrices over Z_m," J. für reine und ange. Math. 238(1969), 85-88.

[103] ______________, "On the enumeration of matrices over finite commutative rings," Amer. Math. Monthly 76(1969), 1019-1023.

[104] L. Carlitz, "Functions and polynomials (mod p^n)," Acta Arith. 9(1964), 67-78.

[105] K. Chew and S. Lawn, "Residually finite rings," Canad. J. Math. 22(1970), 92-101.

[106] W. E. Clark, "A coefficient ring for finite non-commutative rings," Proc. A.M.S. 33(1973), 25-28.

[107] W. E. Clark and D. Drake, "Finite chain rings," Abhandlungen Math. Sem. Univ. Hamburg 39(1973), 147-153.

[108] W. E. Clark and J. Liang, "Enumeration of finite commutative chain rings," J. of Algebra 27(1973), 445-453.

[109] A. Davis, "The Euler-Fermat theorem for matrices," Duke Math. J. 18(1951), 613-617.

[110] R. Davis, "Certain matrix equations over rings of integers," Duke Math. J. 35(1968), 49-60.

[111] K. Eldridge, "Orders for finite noncommutative rings with unity," Amer. Math. Monthly 75(1968), 512-514.

[112] D. Erickson, "Orders for finite non-commutative rings," Amer. Math. Monthly 73(1966), 376-377.

[113] L. Fuller, "A canonical form for matrices over a principal ideal domain modulo m," Canad. J. Math. 7(1955), 54-59.

[114] J. Fisher, "Finite principal ideal rings," preprint.

[115] N. Ganesan, "Properties of rings with a finite number of zero divisors," Math. Annalen 157(1964), 215-218.

[116] __________, "Properties of rings with a finite number of zero divisors," Math. Annalen 161(1965), 241-246.

[117] G. Ganske and B. McDonald, "Finite local rings," Rocky Mountain J. Math. 3(1973), 521-540.

[118] R. Gilmer and J. Mott, "Associative rings of order p^3," preprint.

[119] J. Knopfmacher, "Arithmetical properties of finite rings and algebras, and analytic number theory," J. für reine und ange. Math. 252(1972), 16-43.

[120] ______________, "Arithmetical properties of finite rings and algebras, and analytic number theory, II," J. für reine und ange. Math. 254(1972), 74-99.

[121] ______________, "Arithmetical properties of finite rings and algebras, and analytic number theory, III," J. für reine und ange. Math. 259(1973), 157-170.

[122] K. Koh, "On properties of rings with a finite number of zero divisors," Math. Annalen 171(1967), 79-80.

[123] R. Kruse, "Identities satisfied by a finite ring," preprint.

[124] R. Kruse and D. Price, "Enumerating finite rings," J. London Math. Soc. 2(1970), 149-159.

[125] M. Maxfield, "The order of a matrix under multiplication (modulo m) ," Duke Math. J. 18(1951), 619-621.

[126] B. McDonald, "Enumeration of classes of row equivalent matrices over a principal ideal domain modulo p^m ," Duke Math. J. 37(1970), 163-169.

[127] ___________, "Enumeration of canonical matrices under row equivalence over a principal ideal domain modulo p^n ," Duke Math. J. 38(1970), 393-402.

[128] ___________, "Diagonal equivalence of matrices over a finite commutative ring," J. Comb. Thy. 13(1972), 100-104.

[129] I. Niven, "Fermat's theorem for matrices," Duke Math. J. 15(1948), 823-826.

[130] J. Pomfret, "Similarity of matrices over finite rings," preprint.

[131] R. Raghavendran, "Finite associative rings," Compositio Math. 21(1969), 195-229.

[132] M. Rosen, "Direct products of cyclic groups," Amer. Math. Monthly 73(1966), 55-58.

[133] J. Schneider, "On principal complete local rings," preprint.

[134] D. Singmaster, "Rings of order four," Monthly Problem E1648, Amer. Math. Monthly 71(1964), 918-920.

[135] A. Simis, "Residual finiteness of rings under ring extensions," preprint.

[136] R. Wilson, "On the structure of finite rings," preprint.

[137] ________, "On the structure of finite rings, II," preprint.

[138] ________, "Representations of finite rings," preprint.

[139] R. Wirt, Finite Non-Commutative Local Rings, Thesis: University of Oklahoma, 1972.

[140] R. Wirt and B. McDonald, "The role of the Galois ring in the theory of finite rings," preprint.

G. The Units of a Finite Ring

[141] C. Ayoub, "On the units in certain integral domains," Archiv. der Math. 19(1968), 43-46.

[142] ________, "On finite primary rings and their groups of units," Compositio Math. 21(1969), 247-252.

[143] ________, "On diagrams for Abelian groups," J. Number Thy. 2(1970), 442-458.

[144] ________, "On the group of units of certain rings," J. Number Thy. 4(1972), 383-403.

[145] J. Dennin and D. McQuillan, "A note on the classical groups over semi-local rings," Proc. Royal Irish Academy 68(1969), 1-4.

[146] S. Ditor, "On the group of units of a ring," Amer. Math. Monthly 78(1971), 322-323.

[147] N. Eggert, Some Results on Finite Rings Whose Groups of Units are Abelian, Thesis: University of Colorado, 1970.

[148] K. Eldridge, "On ring structures determined by groups," Proc. A.M.S. 23(1969), 472-477.

[149] K. Eldridge and I. Fischer, "D.C.C. rings with a cyclic group of units," Duke Math. J. 34(1967), 243-248.

[150] ________________________, "Artinian rings with a cyclic quasi-regular group," Duke Math. J. 36(1969), 43-48.

[151] E. Elgethun, "Subgroups of central separable algebras," preprint.

[152] H. Farahat, "The multiplicative groups of a ring," Math. Zeitschr. 87(1965), 378-384.

[153] R. Gilmer, "Finite rings having a cyclic group of units," Amer. J. Math. 85(1963), 447-452.

[154] W. Klingenberg, "Lineare Gruppen über lokalen Ringen," Amer. J. Math. 83(1961), 137-153.

[155] E. Malek, "On the group of units of a finite R-algebra," J. of Algebra 23(1972), 538-552.

[156] W. Nicholson, "Semiperfect rings with Abelian group of units," Pacific J. Math. 49(1973), 191-198.

[157] K. Pearson, "On the units of a modular group ring," Bull. Austral. Math. Soc. 7(1972), 169-182.

[158] K. Pearson and J. Schneider, "Rings with a cyclic group of units," preprint.

[159] ________________________, "On the group of units of a finite commutative ring," preprint.

[160] J. Pomfret and B. McDonald, "Subgroup structure of the general linear group $GL_n(R)$, R a local ring," preprint.

[161] ______________________________, "Automorphisms of $GL_n(R)$, R a local ring," *Trans. A.M.S.* 173(1972), 379-388.

[162] R. Raghavendran, "A class of finite rings," *Compositio Math.* 22(1970), 49-57.

[163] I. Stewart, "Finite rings with a specified group of units," *Math. Zeitschr.* 126(1972), 51-58.

H. The Galois Theory of Artinian Rings

[164] G. Azumaya, "Galois theory for uni-serial rings," *J. Math. Soc. Japan* 1(1949), 130-146.

[165] L. Childs and F. DeMeyer, "On automorphisms of separable algebras," *Pacific J. Math.* 23(1967), 25-34.

[166] C. Curtis, "Quasi-Frobenius rings and Galois theory," *Illinois J. Math.* 3(1959), 134-144.

[167] F. DeMeyer, "Some notes on the general Galois theory of rings," *Osaka J. Math.* 2(1965), 117-127.

[168] M. Harada, "Supplementary results on Galois extensions," *Osaka J. Math.* 2(1965), 343-350.

[169] T. Kanzaki, "On commutator rings and the Galois theory of separable algebras," *Osaka J. Math.* 1(1964), 103-115.

[170] __________, "On Galois algebra over a commutative ring," *Osaka J. Math.* 2(1965), 309-317.

[171] __________, "On Galois extension of rings," *Nagoya Math. J.* 26-27(1966), 43-49.

[172] __________, "A note on Abelian Galois algebra over a commutative ring," Osaka J. Math. 3(1966), 1-6.

[173] H. Kreimer, "On the Galois theory of separable algebras," Pacific J. Math. 34(1970), 729-740.

[174] T. Nakayama, "Galois theory for rings with minimum condition," J. Math. Soc. Japan 1(1949), 203-216.

[175] __________, "Non-normal Galois theory for non-commutative and non-semisimple rings," Canad. J. Math. 3(1951), 208-218.

[176] __________, "Generalized Galois theory for rings with minimum condition," Amer. J. Math. 73(1951), 1-14.

[177] __________, "Generalized Galois theory for rings with minimum condition, II," Amer. J. Math. 77(1955), 1-16.

[178] Y. Takevchi, "On Galois extensions over commutative rings," Osaka J. Math. 2(1965), 137-145.

[179] H. Tominaga, "A note on Galois theory of primary rings," Math. J. Okayama Univ. 8(1958), 117-123.

INDEX